Fixing Physics

not all potentials are created equal

by

John P. & Michael J. Wallace

Casting Analysis Corp.
Weyers Cave, Virginia 24486
www.castinganalysis.com

Made in USA

1st printing July 2024
2nd printing January 2025 with additions.
3rd printing July 2025 with additions.

to

JANICE

Part No. pm5-manual-vol-10-1.23

ISBN-13: 978-0-9986713-9-0

also from Casting Analysis Corp.

DARK MATTER FROM LIGHT
EXTENDING QUANTUM MECHANICS TO NEWTON'S FIRST LAW

THE PRINCIPLES OF MATTER AMENDING QUANTUM MECHANICS

YES VIRGINIA, "QUANTUM MECHANICS CAN BE UNDERSTOOD" 2ND EDITION

YES VIRGINIA, "QUANTUM MECHANICS CAN BE UNDERSTOOD" 3RD EDITION RELEASE IN 2025

Contents

Preface

Joseph Stalin directed his comrades in the late 1940s to infiltrate physics departments in the free world so that no future advances like atomic energy would be missed by the Soviets [1] [2]. By the mid 1970s this subterfuge worked so well with the tenure system that not only physics departments were trashed, but so were the rest of the departments in most universities. These politically selected faculties, taken from a narrow segment of the population ensured their incompetence, which then promoted the flaws of the 1930s already embedded in physics to produce generations of students that halted progress in physics.

A key mistake was made in 1926 by Schrödinger and Dirac when writing down their quantum wave equations. Their incorporation of the non-relativistic energy operator was the culprit [3]. A simple mistake of dropping an additive constant, the particle's self-energy, was never caught. The point missed is that relativity operates at all energy scales, even at rest and there is no correct version of non-relativistic quantum mechanics. By 1932 Dirac realized they were now compounding errors. The mistakes of 1926-1932 were frozen into quantum mechanics.

The research budget cuts in the 1970s at the end of the Vietnam war were one of the best breaks for science and technology in the US partially thwarting the programs of academic physics. Graduates and postdocs had to go out and find employment in great numbers and that powered the semiconductor and biotechnology booms in private enterprise that are still in progress.

The errors in physics did not go unnoticed.

- EINSTEIN'S DISSATISFACTION WITH QUANTUM MECHANICS [4].

- DIRAC'S FEELINGS THAT THE ENTIRE SUBJECT NEEDED TO BE RE-DONE AS HE STATED AT THE END OF THE LAST EDITION OF HIS FAMOUS BOOK QUANTUM MECHANICS [5].

- ENRICO FERMI'S REJECTION OF DYSON'S PERTURBATION METHODS FOR MESONS DERIVED FROM QUANTUM ELECTRODYNAMICS [6].

- POLYKARP KUSCH'S REJECTED THE USE OF SINGULARITIES IN DE-SCRIBING PARTICLES PROMPTED HIM TO BEGIN EXPERIMENTS WITH

ATOMIC HYDROGEN IN 1966 TO SHOW QUANTUM ELECTRODYNAMICS
WAS INVALID.

Enforcement of these modern dogma by the Stalin school of physics was accomplished by censoring any contrary publication through the controlled science press. Over the years this led to a neutering of the university educational system by dumbing down the instruction that spread to secondary education producing a real disaster by leading to a popular theory of man dominating climate change while ignoring geophysical data [7] [8].

Faraday in 1831 found a simple experiment with an iron transformer core where large scale quantum processes were active. Good magnetic induction data has been available for more than 150 years [9] showing major inconsistency with the Faraday-Maxwell form of electromagnetism where errors exceeded 1000%. These low energy physics experiments with quantum fields have continued to the present and are much more revealing than any high energy accelerator experiments.

For a longer work on this subject we use a period hit squad from the 1930s recruited from P.G. Wodehouse's novels to lay waste to modern physics. In that work all the relevant arguments are derived in detail [10]. Bertie and Jeeves had managed to confound the fascist and communists elements in their world and they now can deal with the same zombies in physics. Physics is after all a game best played by amateurs. In this work we are covering the key points about real physical spaces, but for those enthusiasts that would like to get on with physics, there is a complete tool box at the end.

John P. Wallace
Weyers Cave, VA

Michael J. Wallace
Phoenix, AZ

14 July 2025

Chapter 1

Fashionable Ideologies

"If it isn't simple, it isn't physics"

Polykarp Kusch as told to Jacques Barzun [11]

1.1   Simplicity

Simplicity has always been a problem for physics: because there is little context for simplicity. Nature has to resort to simplicity to function, complexity has a high cost in terms of energy and information, both are frugal resources not to be wasted.

The problem is people entangled in organizations love complexity as a cover to survive and for them simplicity is unnerving. Physics embodies simple ideas that have to be suppressed so it can be organized in a political structure as something complex that deserves great expenditures.

Physics advances slowly no matter the amount of monies spent or the popular nonsense in circulation. Currently physics is mired in the mathematics of symmetry that was forced on the subject by mathematicians wanting to embed symmetries into the basic fabric of space. The entire history of physics needs to be understood, not just the recent past. Currently it is unprofessional to cite research more than five years old that has any connection to physics' history.

Avoiding historical citations was the advice cheerfully given by a well known physicist years ago at one of our national labs *" . . . remove all citations beyond five years, as a necessity in order to get a paper published"* [12]. His well tested logic was that editors want things fashionably up to date, while not being seen to be complicit in hiding lies about the questionable foundation of quantum mechanics. A look back into the history of physics is necessary to discover where things went wrong. Obviously this young man was too trusting to see that his contemporaries were hiding things. Checking the history of physics it is easy to find errors, as many are the source of major arguments. After finding some recent errors, it was necessary to move further back in time before Newton's contributions to Pythagoras to gain an early perspective of the subject.

Learning about physics can be a great entertainment. Doing physics is something that is usually not entered into lightly as there are long odds against anyone adding something useful. Physics is not a profession like medicine or civil engineering, it covers essentially everything physical. Its broad range compromises the competence of the practitioners opening the subject to fraud not only by individuals, but by institutions both public and private. The only counter balance are ideas experimentally tested that work: Newton's laws have survived three centuries. Even more demanding is the requirement that models are self consistent across the wide range of physical situations. This is a hard standard to meet. That means man made internal divisions in the subject such as classical, quantum, relativistic, or high energy are false diversions.

The official discovery of quantum mechanics originated in a long term difficulty interpreting the colors of hot electric light filaments. Because there was good data for this problem the subject advanced. The division between the mechanics of Newton's laws and quantum mechanics was bridged by Paul Ehrenfest who showed quantum mechanics was the basis. One advantage in studying physics is that its basic principles make it a compact subject.

The best example of the failure of quantum mechanics is the inability to accurately compute the hydrogen ground state energy [13]. It is essential to work up the energy scale to minimize confusion. Complexity intrudes at higher energy. Working at low energy is an unfashionable approach to what takes place at CERN and the national accelerator laboratories that use the lie that high energy physics is the most fundamental approach. Their inherent experimental roadblock is that their data is sparse with major gaps in what can be measured and then disguised by applying the current model of the day [14]. The object is to gain a profound understanding of one area that will be valid over the entire subject that ends up conveying significant power.

The main barrier to advancing physics is the educational dogma of dividing the subject into separate categories. Nature is not packaged in chapters with neat separations between different sub-categories. Nature comes at you head on in her full glory not expecting you to figure out any of her subtle tricks. Nature cares little for the misleading false rigor of mathematicians in narrow categories. Questions of value are encountered in trying to build and make new things or in trying to understand results from measurements on common materials. It is typically the shame in the failure to understand a common phenomenon that drives an inquiry.

Sir Humphrey Davy warned the young Michael Faraday that doing science is like dealing with a very strict mistress. This is an accurate description and Michael Faraday was blessed by being self-taught and then finished his apprenticeship with an accomplished chemist. His education in mathematics was weak and did not include the use of calculus or analysis. Mathematics is not a strict mistress, but more like a temptress always ready to mislead with a greater frequency than in revealing any truth. Because Faraday's mathematical education was lacking it allowed him to open his imagination to accurately describe the invisible forces of electricity and magnetism.

1.2   The Architects

Quantum mechanics was reported to have appeared as an accidental pregnancy born from the conflicted mind of Max Planck. A Prussian, who did not appreciate statistical uncertainty, found himself in a tight spot having to produce a quick result. He came up with the only thing that worked, the Quanta. Its birth in 1900 after a long and unnoticed pregnancy a century after Young's famous double slit experiment, and two centuries removed from the debate Newton had with a Dutchman on whether light was a particle or a wave. That was the beginning of problems for this unwanted child, because physics was thought by many to be almost totally wrapped up in the 1890s and the Quanta was an unnecessary family addition. The opposition to the quantum characteristics of the photon proposed by Einstein fought by Millikan and vexing Bohr have continued on for more than a hundred years since 1905. Planck's problem with Einstein's view of light and those who followed him really started with Newton and his particle picture of light. This packet of discreet information, not some fuzzy concept, was a concrete observation that obeyed strict energy and momentum conservation. The photon's complete character had not yet been captured in any quantum model and this is one of the mysteries confronting a study of physics.

The subject of quantum mechanics is like an ill kept construction site with much trash thrown about hiding the foundation, which is covered in mud after a rain. Those setting fashion in the sciences in the latter half of the 20th century, mathematicians and people like Richard Feynman, operated like the finishing carpenters, painters, and drapers before the masonry has been set for the basement walls. Physics is an experimental science and so is quantum mechanics, with the tripping point being the word experimental. Those who think one can dream up solutions independent of well thought out experiments, particularly when it comes to the foundation of quantum mechanics, have had their innings with very little to show. Key experiments have been collecting dust for centuries, even though ignored, provide starting material to reexamine the whole of quantum mechanics and the geometry that supports the subject.

1.2.1   Fashionable Physics

Feynman's recorded utterances give a hint about troubles he had ignored in his acceptance lecture for the Nobel prize on 11 December 1965:

" THE CHANCE IS HIGH THAT THE TRUTH LIE IN THE FASHION-ABLE DIRECTION. BUT, ON THE OFF CHANCE THAT IT IS IN ANOTHER DIRECTION – A DIRECTION OBVIOUS FROM AN UNFASHIONABLE VIEW OF FIELD THEORY – WHO WILL FIND IT? ONLY SOMEONE WHO HAS SACRIFICED HIMSELF BY TEACHING HIMSELF QUANTUM ELECTRODY-NAMICS FROM A PECULIAR AND UNFASHIONABLE POINT OF VIEW; ONE THAT HE MAY HAVE TO INVENT FOR HIMSELF "

Experimental science being swept along in a current of fashion is not going to end well. A rule experimentalist have discovered about physics is that there were not multiple ways to explain basic properties: one or none. It is a point of logic that no true proposition requires a qualification. Having any secondary explanation is a qualification indicating the original statement is false. The implication is that fashion may be a terminal distraction, a form of self-delusion, is definitely a true statement.

Mathematical physics has advanced by discovering differential equations that give accurate descriptions of physical phenomenon. It is easier at times to discover an equation than to design an experiment. The hardest part is understanding all that the equations imply about their solutions and the spaces they embed. Learning from experiments was Planck's method of hatching quantum mechanics. To discover an equation for quantum mechanics, while not deriving it from some elementary principles is the way the Schrödinger and Dirac equation appeared. This is a poor way to begin a new subject. The Schrödinger equation was not derived, but simply written down using some weak arguments from classical mechanics [15]. The utility of the Schrödinger equation was compromised even for low energy problems because of the failure to incorporate relativity. A partial derivation of the Schrödinger equation with relativity was recently found [16] and then a complete derivation quickly followed [17]. Planck's black body radiation equation was derived for the expedience of fitting data, but the details describing the photon were not understood. These gaps in the logical connection between description and understanding allowed confusion, and in those gaps fashion rules.

Our problem is where to start in a field rich with defects of logical and experimental inconsistency? The path here will only deal with the basics as there is no need to get tangled in a mathematical nightmare if it can be disposed of as a whole. In table 1.1 are a few of the obstacles that litter physics, however, approaching each one head on with an attitude like Don Quito's will produce frustration. To make progress with physics some knowledge of its history from original sources is essential.

TABLE 1.1: CURRENT QUESTIONABLE FASHIONS IN PHYSICS

QUANTUM MECHANICS IS ABOUT THE SMALL

SINGULARITIES ARE ACCEPTABLE

NEUTRINOS HAVE A REAL MASS

THE ELECTRON IS A POINT CHARGE

SINGLE VIRTUAL PHOTONS EXIST

SYMMETRIES ARE PRIMARY NOT SECONDARY PROPERTIES

1.3   History

> *The Past is such a curious Creature*
> *To look her in the Face*
> *A Transport may reward us*
> *Or a Disgrace —*
>
> *Unarmed if any meet her*
> *I charge him fly*
> *Her rusty Ammunition*
> *Might yet reply*

EMILY DICKINSON

First, a short background history with an effort to only use original sources of those who actually did the work. Derived histories from secondary sources on 20th century physics can reflect more about the historian than the original work. Einstein, Dirac, and Fermi are very clear in their writings about what they covered and what they ignored. The other important features about papers from the original sources before 1940 is that they are highly readable unlike the transformation that took place in the 1960s and 1970s when physics papers became obscure, even before the post modernist movement championed total obfuscation.

History of physics sets the stage for what all the fuss is about with quantum mechanics. Understanding experiments before 1900 is necessary, because the unofficial discovery of the subject goes far back in time. In 1927 a very wise and stubborn man, Paul Ehrenfest, showed the principal law of classical mechanics, Newton's second law of motion, could be derived from quantum mechanics, but not the other way around [18]. Neither he nor his friends at the time realized the importance of what he had done. This was a major hint that the rest of physics concerning matter and fields should find its origin in quantum mechanics.

1.3.1   Newton

One can ask the question "What has Newton to do with the foundation of quantum mechanics?" The answer is just about everything because he did so much that touched on the basics of quantum mechanics. Both on the physics of mass and his calculus exposing limits to spatial measurements in what he called fluxions. Newton had a problem when it came to publishing the *Principia*, whether or not to include the calculus [19]. He did not include the calculus and stuck with geometry. The calculus would have opened questions about space and how it could be divided. These questions could not be answered simply, so they were avoided. It took a couple of centuries for the mathematicians to rigorously treat and expose the problems of understanding continuum spaces. Newton's assumptions about the properties of space were useful for his mechanics but could not be experimentally verified in the 17th century. It did open the question of physical space as an important concept to be understood and attracted the likes of Gauss and Riemann to consider the problem.

1.4   Early Experiments

The physics of Copernicus, Kepler, Galileo, Newton and even Einstein was dynamics, things moving, computing orbits. Physicists became skilled at solving dynamical problems, and this momentum moved through the discovery of quantum mechanics. Bohr and his quantized orbits and Wolfgang Pauli were fascinated with Kepler's view of the universe and his work on orbits. This fascination with Kepler's work in the 20th century is found in Pauli's work and those of his students [20]. Feynman's method of summing all possible paths for particles and fields is just an off shoot of these 17th century studies in dynamics. Dynamics rather than static matter was of greater interest to physicists because of planetary astronomy and the need for more accurate artillery tables. Static properties presented a more difficult challenge. The other main development in physics was the study of fields associated with Faraday and Maxwell. It appeared almost as a separate discipline compared to the dynamics of classical physics. These two vastly different parts of physics can only merge when quantum mechanics is properly generated. The major challenge now for quantum mechanics turns out to be what it can tells us about static material properties?

The problem of a quantum object at rest lacked a formal description until recently [16]. The relative state of rest is only defined for a particle with mass, since fields move at the speed of light they don't normally come to rest. The details of a stable light nucleus are still a mystery to us. Understanding the rest state has not gotten very far in physics because if a particle is not moving there is little that can be measured. The rest state was important for Newton as the concept is contained in his first law of motion. The mistake made in studying a particle at rest is assuming relativity is no longer important and that is a trap almost everyone has fallen into. A particle's structure is dependent on relativity either at rest or in motion. The other half of the coin are the fields that are always in motion, light and neutrinos, the problem for them is also simply stated. If you think of photon as a spherical wave moving outward uniformly as a shell of radiation, for this is the mathematical description of a photon, how can it possess a directed linear momentum? That is not far removed from the puzzle about static matter. Fortunately, there is good experimental data, some of it recent, to help untangle both of these problems simultaneously.

Newton and Huygen were battling over the quantum nature of light in the late 17th century. Michael Faraday believed that the electrical forces were connected to the gravitational forces and tried experimentally to find this connection. Relativity had not yet been discovered and this barred them from asking serious questions of what is matter. Science had advanced a little quicker in the studies of light and radiations. The experimental work on light, electricity, and magnetism led at the beginning of the 20th century to the discovery of quantum mechanics.

1.4.1 Thomas Young, 1801

The first major large scale experiment in quantum mechanics was carried out by Thomas Young and it predates Faraday's induction experiment by a few decades. His double slit diffraction experiment produced an interference pattern. When done with an individual photon, electron, or bucky ball one at a time the experiment produces the same result as obtained when done with an intense source casting a pattern of interference fringes of light and dark. The question it raised was how does a particle interfere with itself, or stated another way how can a particle be in two places at once? The mask with the slits divides the particle paths into two domains and the question to ask is: Why are these two domains being treated the same with respect to what is flowing to the mask? The double slit experiment exposes both an ignorance of the structure of the particles moving through the slit and the physical space they occupy. A paradox in understanding is the result.

The answer to this paradox requires considering how the radiation or particle field is defined in its own frame of reference. Introducing an obstacle, the double slit, into this experiment highlights a part of quantum mechanics where field behavior of the particle stops being a paradox when the space is treated as a whole. The question then becomes: How is the isolation of particle or field even accomplished?

The double slit experiment is often the first problem seen in the text books on quantum mechanics and it is presented as a paradox compared to classical mechanics where particles are modeled as ball bearings. Unfortunately, the lesson usually stops there and not opened for discussion as to what would be required to produce the result. The present description of quantum mechanics is not adequate to explain the results.

The interference experiment when done with a single photon or electron in the device at a time is an open question. How can a particle interfere with itself? The photon is a field and that can almost be justified, but an electron is a particle, how can it be in two places at once? To actually solve this problem a

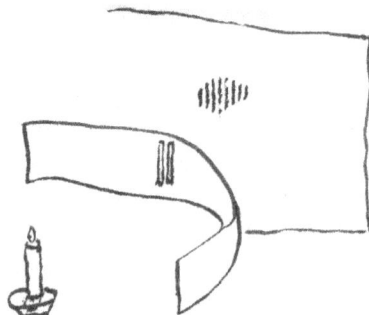

Figure 1.1: **The diffraction pattern in a Young double slit experiment depends heavily on the wave front description of a spherical wave for the photon intercepting the two slits [21]. A spherical expanding wave front is the description of the photon's field in its own frame of reference [17].**

frame of reference associated with the particle or field must be introduced satisfying the constraints of relativity. What results is a better understanding of what is required to be an allowed quantum entity such as an electron, a neutrino, or a photon. How is a quantum object constructed so it is capable of carrying the information about its own structure? These are the set of questions that have been recently answered [16] [22]. Unfortunately, the majority committee that does quantum mechanics today would say such questions are not allowed under the Copenhagen interpretation of quantum mechanics. The rules from Copenhagen are going to be abused in the extreme.

1.4.2   Michael Faraday, 1831

Michael Faraday on the 29 of August 1831 using an annealed 6" diameter forged ring of iron as a transformer core, measured an induced electromagnetic impulse passed between two circuits. Closing a switch to a battery connected to an insulated copper wire winding on one side of the torus generated a signal in a secondary winding on the other side connected to a galvanometer. Faraday's experiment and the Faraday-Maxwell equation for the description of induction is accurate in free space, but when well annealed iron is used as a coupling agent the problem becomes one of the more interesting large scale quantum mechanical problems that exist [23] [24]. Toroidal transformer cores were used by Henry Roland in the late 1870s to study the dynamics ferromagnetic permeability of different ferromagnetic metals and alloys. The dynamic measurements of permeability in pure iron continued for the next 50 years with improved iron purity and at elevated temperatures still producing the unexplained large values. The permeability measuring technique was then dropped at Bell Labs for a quasi-static one less theoretically troublesome.

Faraday's dynamic permeability test would often result in values greater than a million and did not match smaller values produced in quasi-static permeability measurements. There was a problem with the time dependent description of the field within iron. This effect showed up with induction heating losses below the Curie point in magnetically soft steels [25]. The Curie point is the temperature below which iron is a ferromagnetic material. We discovered this permeability measurement problem as an inability for any value of the permeability to characterize the reflection of an electromagnetic field from annealed steels, when computing the reflection from steel using Maxwell's equations independent of any form of the $B = \mu H$ relation [26] [27]. This relation describes the strength of the polarization of a magnetic material by a known external magnetic field. Any form of the material equations for magnetization simply did not work. In addition, simple application of quantum mechanics could not untangle the process that was taking place, because the dynamics of what was taking place were unknown. A failure in Maxwell's equation to describe such a simple experiment indicated a defect in our understanding of both quantum mechanics and electromagnetic theory.

We know a great deal more about the metallurgy of iron than we know about quantum mechanics. Iron has been studied for 3.5 millennium as people had been making notes about

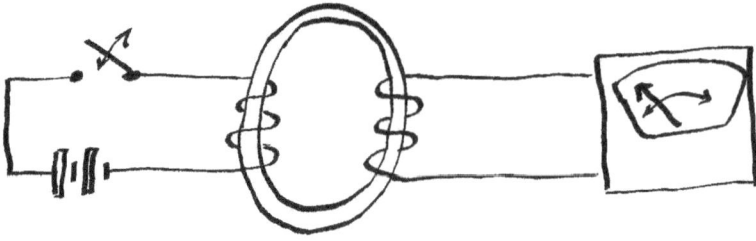

Figure 1.2: **Original 1831 induction experiment that produces a transient deflection when the switch is closed. The coupling element is a well annealed iron ring 6 inches in diameter.**

the metallurgy of steel. The failure in understanding dynamic ferromagnetism originates with a defect in the foundation of quantum mechanics. The defect was the poor way in which relativity has been integrated into quantum mechanics [28] [16]. For Faraday the quantum effects in his transformer core were not a problem, only an advantage for boosting the signal.

1.4.3   James C. Maxwell 1860s

Maxwell is best known for introducing a self-contained theory explaining electricity and magnetism. His conception of fields was built upon some very mechanical models that included elements that were anything but abstract. These mechanical models disappeared as his many equations were reduced by modern notation to four equations by Heaviside. The only remnant of the mechanical legacy that survived was the concept of an aether to relieve their concerns how fields propagate in empty space. Most think the aether was swept away by the Michelson-Morley experiment of 1887 with the curious result of the independence of the velocity of light with respect to relative motion of the measurement frame. When Maxwell did speculate on these matters, he put it at the very end of the second volume of his major work [29].

The concept of point mass and point charge were bigger problems. These two concepts evolved from experience that atoms are very small. The point charge was a major problem for electromagnetic theory, mathematical singularities, as the concept did not clarify the source of the static electric field. After Maxwell, before any of this was resolved 34 years passed when the quantum nature of light needed to be explained because of the black body radiation data. The confusion for physicists was compounding with the discoveries of this period.

A second area where Maxwell worked was statistical mechanics of an ideal gas and how probabilities enter naturally when there are innumerable interaction events. His name is linked with Ludwig Boltzmann whose interest was providing a foundation for thermodynamics. Statistics and fields are what quantum mechanics is about and these were Maxwell's specialties, unfortunately he died a young man or else quantum mechanics might of started earlier than 1900.

The wave equations derived by Maxwell for the electric and magnetic fields moving at the speed of light was a major advance in understanding fields. The success of electromagnetic theory by encompassing two different field sources, static and dynamic fields, made it a superior theory. Quantum mechanics had not integrated Maxwell's work successfully. What limited both theories was extending them to accurately describe active material participation [30]. A good understanding of matter was lacking.

1.4.4   Four Experiments

The significance of the 1887 Michelson-Morley experiment for quantum mechanics is large: uniformity of space, no preferred directions, and space as a void. The experiment supports the concept of the independence of the velocity of light in any frame of reference in which the measurement is taken. That was an outstanding result. The light which is in the interferometer establishes an absolute frame of reference independent of the

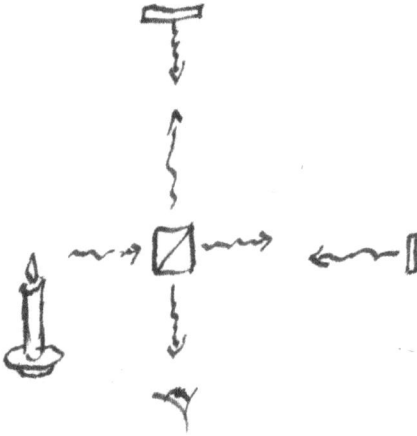

Figure 1.3: **The 1887 interferometer's interference pattern of dark and light bars for the observer reflect the same geometry of the spherical photon wave front as found for the double slit experiment.**

laboratory, the solar system, and even distance stars. This is a concept that had no meaning in terms of the classical physics of the day. Associated optical experiments show that light can be superimposed with no net interaction. The question that was not answered was what does this imply about the description of light within the interferometer? The light is neither behaving in the Newtonian particle model nor as a simple wave model in a medium. The light seem to move with its own personal propagating medium attached to itself, independent of the rest of the universe. This forbidding isolation is required for superposition, but makes no connection to the contemporary physics of the period.

The notoriety of Einstein solving the puzzle of the photoelectric effect was also a beginning for all scattering experiments. The experiment was a precursor to all future accelerator experiments. It was an experimental method for detecting the energy threshold of either liberating a bound particle or creating a particle and nothing more. One knew what was entering the scattering center and what came out, however, any information about the actual excitation process was never available. It was the *"nothing more"* component that was lost on physicists, even though in the mid 1960s Mark Kac pointed out this limitation [31]. People made up stories about what actually took place during the excitation that eventually led to theories like the Standard Model of particle physics.

The best experiment of the period that constrained attempts to violate the conservation of energy and allowed quantum mechanics to get a better start was the Compton scattering experiment of 1923. Where electrons and x-rays scattering off of each other gave a measure of the electron's scattering volume. The secondary result showed that both momentum and energy are conserved. Defining a volume for the electron was an experimental landmark, signaling things needed to be fixed in both potential theory of electromagnetism and quantum mechanics by marking the end of the point charge. Unfortunately, the point charge as a concept did not die even with its infinite energy. The ferrous induction experiment by Faraday, Young's double slit experiment, the Michelson-Morley experiment, and the Compton result became museum pieces and could be safely ignored as they were no longer fashionable. The fashion of violating energy conservation ruled out some very good experiments and slowed the development of physics.

1.4.5   20th Century

Relativity gained a shaky foothold in 1928 in quantum mechanics with the Dirac equation making some major compromises. Dirac's efforts to make quantum mechanics relativistic poisoned the subject with two simple mistakes: using a non-relativistic energy operator [32] and taking the square root of the relativistic conservation expression. Mathematicians got to work sealing off the subject with a single mathematical space,

Hilbert space, formalized by J. Von Neumann in 1932 [33].

Dirac's misstep was trying to force relativistic quantum mechanics to fit a first order linear expression, $E \sim p$, that worked for radiation. This error was compounded by Einstein's reluctance to deal with his second order relativistic conservation of energy relation suppressing any further understanding of quantum behavior. They were both use to using linear energy relations of thermodynamics and statistical mechanics and were not eager to embrace a quadratic relationship for energy conservation. The problems for Einstein and Dirac were three fold in applying the quadratic energy relation: first they did not have experimental accelerator data verifying the result, 2) potentials were not treated in the relation, and 3) there was the open question of how to explain the orthogonality between the self-energy and the kinetic energy. This collection of problems forced them to use a linear energy relationship for quantum mechanics.

$$E = \hbar\omega \ field \quad \& \quad E^2 = p^2 c^2 + (m_0 c^2)^2 \ particle \qquad (1.1)$$

The first equation due to Planck, the energy of the field, is proportional to the product of Planck's constant divided by 2π which is written as $\hbar$ times the frequency times 2π, which is ω for the massless field. The relationship works because radiation whether electromagnetic or in the form of a neutrino has to be dealt with as a whole, not something that can be fractured. The photon or neutrino either exists or do not exist. The second relation from Einstein for a particle with a rest mass, m_0, and a momentum, p where c is the speed of light the quadratic relationship automatically introduces an orthogonality between the kinetic energy and the self-energy of a particle. It is how this orthogonality is generated physically that needs to be understood.

The expression for the energy, E for light, is a first order equation and for a particle with mass the energy dependence is a more complex quadratic equation. These are the two most important relations in physics. These two empirically verified expressions will generate the allowed particles and fields along with the forces [16] [22]. These two equations appear simple and almost trivial, they conceal a great deal. They can easily seduce one into thinking there is not much to understand. The one defect they both seem to possess is there are no potential energies expressed. There was a reason for this lapse as not all potential fields can be treated the same and that detail has long been avoided.

The Copenhagen school of quantum mechanics left a set of unanswered questions enshrining an ignorance that was supported and defended by the editors of the physics journals [34]. The main problem was how to go about deriving the Schrödinger equation or matrix mechanics? What is the quantum description of the particle itself, rather than a description of its dynamic behavior? Why is there a statistical basis for quantum mechanics? These questions are coupled because there was a problem with the electron, which was assumed to be a point charge and as such would have an infinite self-energy from its electrostatic field. This is more energy than needed to create the electron in the first place. This question cannot be answered until there was a physically tested description of the electron that has an electrostatic field with a finite self-energy [35].

As a group, physicists, look like they have much in common with the ostrich and this is accomplished by simply not looking back. The American Institute of Physics prohibition of older citations gave young physicists the cover they needed to avoid these old experiments: double-slit, Compton, Michelson-Morley, and Faraday induction data. Also, the problems are hard and take generations to solve, even though they are known to exist they are not immediate or urgent and might be safely ignored. In the later half of the 20th century journal editors were paid to be guardians of programs that were paying for the papers being generated. In some cases they championed certain interpretation of quantum mechanics with selected data and they rejected papers that uncovered problems with the foundation of the subject. The fault with this logic is that problems are not solved, but the vacuum they created are filled with nonsense. It is an either or proposition that cannot be escaped, leading to progress or a wasteland.

Einstein took Planck's quantum and carried on his examination into the mid 1920s where he assumed special relativity was complete. The Schrödinger and Dirac equations producing similar but slightly different results for the hydrogen atom's ground state along with the derivation of the magnetic moment of the electron from the Dirac equation satisfied many. They concluded that the basic theory was complete with the integration of relativity. Quantum mechanics for a hydrogen atom worked almost too well with small deviations from experiment. The theory that followed, quantum electrodynamics, continued to employ singularities in the electron's potential to make their final corrections. That should have been a warning, no matter what numbers the models were producing, there was still a problem with any physical understanding.

1.5   1932 and Politics

". . . PROBLEMS IN RADIATION THEORY WHICH DO NOT INVOLVE THE STRUCTURE OF THE ELECTRON HAVE THEIR SATISFACTORY EXPLANATION."

ENRICO FERMI 1932 [36]

The reservation quoted above ending Fermi's 1932 paper *"Quantum Theory of Radiation"* was the standard document that taught most theorists field theory. The Fermi quote was a clear statement that particle structures were a major unknown along with the structure of fields. It predated Dirac's more general ending admonition that quantum mechanics should be scraped and be redone at the end of his famous text *Quantum Mechanics* [5]. Things did not improve over the years, especially with the advent of quantum electrodynamics as the work was trapped by the classical Lagrangian that Dirac for a short time thought might be useful. Harmonic oscillator model overused by Julian Schwinger in his work on quantum electrodynamics followed directly from Fermi's 1932 paper. The photon field is a three dimensional spatial structure that is a transient propagating wave front and has almost nothing geometrically in common with a one dimensional harmonic oscillator over used by Fermi and Schwinger. The linear harmonic oscillator is an approximation of mechanical oscillators found in nature that are almost never truly linear.

Von Neumann's major work, *Mathematical Foundations of Quantum Mechanics*, had difficulties with applying special relativity. Fermi avoided problems with special relativity by dealing with massless fields. These works helped their two authors leave Europe, but disguised questions that slowed progress as politics played a significant negative role. Neither of these two gentlemen went back after WWII to revisit the fundamental issues of quantum mechanics.

1.5.1   Engineers by Practice and Education

Einstein's Swiss patent office apprenticeship and Dirac's education as an electrical engineer strongly influenced their future work. Most patent applications are for mechanical designs and Einstein did a great job on relativity and its kinetic basis, less so dealing with potentials and the quadratic conservation of energy. Maxwell also found this road block with potentials. When Einstein did explore the 5 dimensional work of Kaluza it set a precedent for others to follow for the next century. This was a precursor to what became string theory with many more dimensions, while Einstein was being dismissed as a spent force.

Dirac was a master of dealing with the operational calculus put forward by Heaviside where singularities were commonly dealt with in the development of telegraphy. This comfort with singularities is evident in his creation of the Dirac delta function and a paper of 1931 where he explores what he thinks are physical singularities in elementary fields [37]. The exploitation of singularities became an industry starting with quantum electrodynamics and the work of Gell-Mann and Low [38] then with Penrose and Hawkings [39]. It was a starting point where initially it was thought the perturbation techniques could correct any poor initial assumptions. This turned out to be a false assumption and these models were not repairable.

The influence of Einstein and Dirac cannot be underestimated through their two books. Their greatest error was in not properly exploiting the relativistic energy conservation relation to integrate quantum mechanics with relativity that would have eliminated both the use of singularities and higher dimensional spaces.

1.5.2   Effect of Wars – Censorship

Both the first and second world wars were a disaster for theoretical advancement in physics even though they drove advancements in experimental physics, particularly in nuclear physics. General relativity was produced during the first world war, but the problem was special relativity still needed significant work that had been totally ignored. The wars wiped out debates on the basic conflicts in physics and these were forgotten by the end of the wars. Special relativity had not resolved how different potentials had to be treated in the relativistic conservation of energy relation. Here the development of general relativity disguised the problem, as the gravitational potential is very different than the electrostatic potential. In this confusion the problem was simply dropped. This misstep was so significant that it impeded any future understanding of quantum mechanics [13]. It turns out the two different potentials enter the relativistic conservation of energy equation in different ways.

The debate between Dirac and Pauli about the validity of his first order relativistic equation for quantum mechanics ceased with the rise of the troubles in Europe before WWII and the experimental discovery of the positron. This debate was never resumed following the war and permitted a narrow view to develop around small corrections to quantum mechanics by quantum electrodynamics, where major corrections to the subject were required. To paper over these basic problems historians made cult figures of the principals who developed quantum mechanics and quantum electrodynamics.

The pre-WWII state of quantum mechanics was then canonized after the war. Effectively, this was censorship that Bohr and Heisenberg took advantage of in championing a form of quantum mechanics that was nothing more than a black box that could not be interrogated. The practice of censorship did not die out, but is present today when challenges to the technical merit of major experimental programs are submitted for publications. They are censored as they would be embarrassments to some countries' national pride. Protection of failures in theoretical physics had now been extended to experimental physics.

The practice of censorship is currently active as we discovered when trying to point out some basic mistakes of the 1920s. The difficulty in publishing on the foundation of quantum mechanics in physics journals are no longer stopped by academic prejudice of what was assumed to be complete, but by policies of state that are followed by subsidized editors on most major journals. The current editorial game is to take what is good for their masters use while barring challenges from the public. This assault by the science press helped to produce generations of eunuchs to husband some artificial form of science.

1.5.3   Post World War II

The Lamb Shift is a measure of the energy difference between the atomic $2S$ state and $2P$ state of hydrogen. According to the Dirac/Schrödinger theory the energy of these two atomic states of the electron sitting above the $1S$ ground state should be identical. The measured value showed a difference between states equal to the energy carried by a $\sim 1\,GHz$ microwave photon. The second post WWII experiment was the measurement of the anomalous magnetic moment of the electron by P. Kusch and H. Foley. These two very small incremental pieces of data kicked off a revolution in theoretical physics by employing perturbation sums that obscured the physical source of the measured values [40] [41]. The larger discrepancy between the computed and measured ground state energy of hydrogen was overlooked.

Revolutions are not usually carried out with a light touch and with Hans Bethe leading the way to quantum electrodynamics by using the old point charge and point mass there was no light touch. This was unfortunate, because when solving problems the best results usually come when a very light touch is applied, since one is never really sure what else may be disturbed. This is the same technique required in doing a good comic novel or play exemplified by P.G. Wodehouse. Doing physics for understanding suffers in revolutions, usually there are details buried before the revolution is complete.

The quantum revolution was produced by a series of events: Planck and Einstein's contributions before 1910, Bohr atom, Sommerfield extensions, quantum mechanics of Heisenberg and Schrodinger, Dirac's stab at including special relativity, quantum electrodynamics, and finally the quantum field theories at high energy that morphed into the standard model. This evolution was only the first draft.

The problem with this progression is that difficulties A. Einstein identified in the first stage of the development of the quantum idea were never cleared up. Particularly, what is the quantum basis for radiation and how is it connected to Maxwell's equations? Fermi tried to reconcile this problem with his famous paper on radiation [36] but failed because it could not explain matter's connection to radiation.

In the late 1940s when quantum electrodynamics was introduced it again forced a problematic issue by using the point charge. There were experimentalist who believe the form of quantum mechanics being sold with a point charge was defective along with the non-uniqueness of the perturbation methods. Non-uniqueness and the spirit of the 1920s would be transcribed by Cole Porter later in the 1930s into a musical, *Anything Goes*. This meant you could produce almost any number for an answer to a problem with a little creative fixing of the books. The opposition included Einstein, Dirac, Fermi, and Gamow who thought that primitive physical fields and elementary matter should be described by an exact and unique theory. Those supporting the fiction of quantum electrodynamics were H. Bethe, F. Dyson, J. Wheeler, and R. Oppenheimer, who were all fans of the abstract.

There was an additional problem with the infinite energies generated in quantum electrodynamics by assuming point particles and point charges. This difficulty brought a larger problem as it came with a non-physical geometry without asking the question what is the geometry of a physical space? Instead of things becoming clearer, the tale became more complex in the 1950s and 1960s. High energy models were split off from the rest of physics on their own weak foundations. This resulted in more speculative theories, essentially compounding errors. Any good theory should be completely self-consistent with the rest of what is known about physics, rather than being separated as a high and low energy theory [14].

These new theories were slowly accepted and protected by their complexity and the backing from private and public philanthropies. They were so detailed, that one could almost forget they were only approximations built on a poor mathematical foundation with a complex set of assumptions. The reason the new approach was tolerated is that the energy levels of the hydrogen atom are not very dependent on the details of the potential at small values of r [13]. The modeled discrepancies with experiment were small and ignored till after the second world war [35]. Small discrepancies are a problem in general because there are many ways to produce a small correction and very few ways to test if the methods are correct.

Problems created by the Lamb shift and the electron's magnetic moment precipitated other experiments designed to clear up problems created by theorist using perturbation theory. One proposed experiment was to measure the ionization energy of single electron atom, hydrogen accurately. This was an idea of P. Kusch who previously took the original anomalous magnetic moment data and was unhappy with quantum electrodynamic [11]. He felt there was a fundamental problem that had crept into quantum mechanics after 1929 and it was not resolved by quantum electrodynamics. Again political forces interfered and his work was postponed for another 60 years.

As time passed, non-unique calculation from quantum electrodynamics were accepted with their singular sources. The reason was the quality of physics education had rapidly deteriorated in the 1960s and 70s caused by a tenure system enamored with totalitarian ideologies. The new faculties selected for their politics appeared to be both ignorant and not interested in the debates of the 1930s.

The logical flaw in quantum electrodynamics is that it justified the use of the singularity in the electron's potential. These methods then migrated into applications in different fields such as astrophysics, cosmology, and high energy physics. Quantum electrodynamics was not done with a light touch. This was a tragedy because it did not put a halt to the theoretical rot that began in the 1930s. At the same time high energy physics was taking off in the 1950s and early 1960s with the discovery of many new particles and high energy models were developed to explain what was being observed. These new high energy models created a divide between quantum mechanics done at high and low energy where there should have been no gap. With no real challenges to quantum electrodynamics, the new high energy modeling work continued to use the techniques of quantum electrodynamics.

1.6 Philanthropy and Committees

Carlsburg Brewery, Alexander Flexner, Joseph Stalin, Freeman Dyson, Robert Maxwell, Robert Oppenheimer, Oxford University Press, DoE, NSF, CERN, Simons Foundations, and a host of minor potentates have all come out of the woodwork in order to maintain some control over the direction of physics research. The philanthropies introduced and maintained ideological biases that have little to do with experimental problems that need to be solved. The dangers posed by philanthropies and governments to research are detailed by Jacques Barzun in his work *The House of Intellect* [11].

TABLE 1.2: **20TH CENTURY COMMITTEES DOING PHYSICS LET A FEW BASIC ERRORS SLIDE BY LEADING TO CHAIN OF THOUGHT THAT STARTS WITH "FOR THE WANT OF A NAIL".**

COMMITTEE	PROBLEMS
SCHRÖDINGER & DIRAC EQUATIONS VALID	NOT PROPERLY DERIVED [5] [15]
COPENHAGEN SCHOOL	A LIMIT TO QUESTIONS [42]
QUANTUM ELECTRODYNAMICS	ELECTRON IS A POINT [43]
STANDARD MODEL	ENERGY CONSERVATION INCORRECT [44] [26] [16]
MASSIVE NEUTRINOS	NON-PHYSICAL [45]
GAUGE THEORIES	ARTIFICIAL FIELDS
LOW SYMMETRIES ARE FUNDAMENTAL	ARE RESULTS NOT A START [46]

Henry Raines, a long time union leader, noted syndicates and unions evolve to operate at the level of their basest member in order to protect them. It is a behavior that also applies to research institutes when they circle the wagons. That is why the framers of the US constitution went to so much trouble to separate powers because they knew what a group of stinkers can do. No matter how nice a benefactor is, there will be expectations of some form of duty.

Committees don't have a good record of pushing physics forward. Some committees having had success in an engineering project think that can be carried over to physics and that is a delusion. The difficulties with unknown physics is finding the language for expressing an idea that does not yet exist. In the past it was individuals who moved things forward and for good reason: no initial necessity for a generating a new vocabulary to solve the problem was required. That language has to be invented. How can a committee invent a language when they would rather use a language they already understand, thus guaranteeing failure?

Historically, interesting problems in physics were only tackled by individuals, because in the course of their work they would invent the language for expressing the solution. Michael Faraday employed William Whewell to come up with the vocabulary for electrochemistry, electricity, and magnetism. He knew without the descriptive words no one would comprehend the character of the discoveries. The problem of expressing new ideas is why new physics has been done by individuals. They do not have to put their thoughts into language for others until it develops to a stage where this new idea is realized.

The group working with Neils Bohr version of quantum mechanics used the statistical basis of quantum mechanics to bar any question about structural details [42] [34]. This fallacious position acts as an effective order to cease and desist for any inquiry. Stating that all that could be known about a subject is now known is rather a good indicator, that not very much at all is understood. Nature is much more inventive so its wise to expect surprises.

Simple mistakes were a problem for all these early quantum enthusiasts. While trying to write down a physical form quantum mechanics, many ignored energy conservation. There are many ways to violate the conservation of energy. The first major blunder made was not using the relativistic representation of a single particle energy conservation as the necessary starting point. People became so desperate to generate a working form of quantum mechanics they considered dropping energy conservation all together for any individual process. This was a proposal floated by J. Slater, N. Bohr and H. Kramers in about

1924 and it did not get very far because of some good experimental work on the Compton effect [47].

The modern versions of research institutes, the national labs, IASs, and CERN have shown themselves to be enterprises designed to promote their own existence as employment facilities for cadres of drones, rather than bothering with the solutions of fundamental problems. The fallacy in the utility of institutes of advance physics was detected in the 1950s and 1960s by two gentlemen P. Kusch [48] and J. Barzun, see Section 4.0.1.

1.7 Data

TABLE 1.3: EXPERIMENTAL LOOSE ENDS.

YEAR	EXPERIMENT	REQUIRES
1801	DOUBLE SLIT	PARTICLE/FIELD STRUCTURE
1831	INDUCTION IN IRON	BOSON DESCRIPTION & BEC
1887	MICHELSON-MORLEY	SELF-REFERENCE FRAME
1947	PION	PROTON STABLITY
1966	GROUND STATE OF HYDROGEN	FIX SPECIAL RELATIVITY
2016	SOLAR NEUTRINO DEFICIT	MASSLESS NEUTRINO

There is a great deal we don't understand about how some simple experiments work. Paul Ehrenfest who understood that nature was so tightly interconnected that one can tell if your work was on the wrong tract if it is not consistent in verifying a wide range of physical measurements. Consistency is an argument that was forcefully made by Michael Faraday. Ehrenfest's reasoning involves experimental data. This interconnection between material properties is not easily studied as a group. Usually, a particular property is investigated one at a time and a body of work is built up around each one in isolation. From a teacher's point of view that looks like a good idea so things can be separated and taught individually. Separating areas of learning tends to suppress conflicts between descriptions that would expose erroneous assumptions. Isolating ideas opens the door to making mistakes.

Experiments are good tools to sort out these questions of how properties are connected. If one is careful in designing an experiment that can be modeled by a simple set of equations, with no free parameters, then a test can be made to see if the description is true to the experiment. The simplest bound state problem that needs to be well understood is the energy of the hydrogen ground state.

There is nothing like good data to clear up problems. What is found in physics is a series of experiments extending back more than 300 years that have not been properly dealt with. The chain of mistakes in reasoning after the first third of the 20th century shut down progress in quantum understanding and have to be removed. Having a list of unsolved problems one might think of starting on the diffraction work of Young. To unravel this tangled mess requires picking a starting point far enough back in time, the fourth century BC. The first problem to be researched is the space we occupy.

Questions

1) Special relativity as introduced by Einstein required two things: a ruler and a clock. If you are in an empty part of the universe with only a few particles where are these gadgets going to come from to confirm special relativity?

2) Can a Euclidean point, line, or a regular solid be constructed in our physical space and please explain?

3) In 1878 George Cantor published a paper where he discovered that in the continuum, where there is no limit to the point density on a line, surface, or in a volume, the concept of dimensions was nothing in his continuum but a set of indexes that could be arranged in any convenient way. Why might this be considered a patch of quick sand? What other traps for the unwary did the continuum supply? [49]

4) A common method used in quantum mechanics starts with something called the vacuum state, which is then populated. Is there a logical problem with this starting point?

5) Lattices (graph paper) are popular as a starting point for theorists. Explain the real assumptions imposed when using a lattice as a starting point to investigate spaces.

6) We have a conservation law for energy, could there be one for space?

Chapter 2

Spaces

2.1 Mathematics is a Luxury

Information, in its different forms, has always been a puzzle and appears to be tied to the physics of matter at its origin. To create persistent information the physical space might have to be partitioned or reduced to a form different from the mathematical space of Euclid. The Euclidean space most are familiar with from geometry purports to give a common sense view of space. Common sense is only common if it can be verified by experiment and not much of Euclid's basic constructions actually exist in nature, as they are idealizations. Euclid being of the textbook writing mind set did not bother to include the debates on the divisibility and the indivisibility of space and matter. What is the most efficient way that space can be made to contain information? Here one might take a hint from biology of the cell and create a partition, a cell wall, to define a region of stored information. The cell shows us what the cell wants to show and we are not privileged to see inside the cell unless it is to be destroyed. Mathematicians like to define multiple dimensional spaces, even infinite dimensional spaces where any

31

number of bits can be safely stuffed. The problem is an external structure is introduced that is in excess to the information stored. However, if our universe had only a single particle a partition of space with a boundary into two regions might seem to be the easiest way to store the particle's data, but it might not be the most efficient method.

There is an implied assumption that crops up when partitioning space. The assumption is a boundary must be introduced. Mathematicians have wrestled with this boundary problem on a simple line segment. The confusion over using the greater than, >, or the greater than and equals symbol, ≥, to define a region on a line posses a problem. That last dangling point is an ambiguity requiring a choice between two symbols needs to be avoided in the physical world.

The idea of needing a bounded volume walled off from the rest of space to store data is not particularly original and it is also an incomplete concept. If you had the ability to range over all space and go through the boundary and explore the details within the boundary, then the data there could exist independent of the boundary. That type of boundary is just a convenient way of labeling things and not essential. The ability to defeat a true boundary at will is not physically realistic if the boundary is a necessity. If there is limited or no access to the interior of the bound region then the concept of a partition is on firmer ground. A real barrier to invasion makes a strong case for holding data within a bounded region being available only as a partially visible external property where the contained volume of space generates their external characteristics. So instead of a library of isolated lists of information, the primitive information is generated by the properties of the bounded volume. To the world the bounded space produces a set of basic properties via structure and not as a ledger of stored facts.

Many physicists think the quantum picture of an electron is a structure-less point with a mass, charge, angular momentum, and magnetic moment assigned as a list of attachments to the electron. This sorry state of understanding is so non-physical that it does not fit with any architectural origin of information. Basic properties should be generated from the structure of the particle where these static properties of matter can then form an alphabet for expressing abstract information.

Primitive data and primitive spaces are often encapsulated. Its rather a common feature in nature. It is a basic mechanism associated with defining information itself. An example of encapsulation is the atomic principle validated by J. C. Maxwell and L. Boltzmann in their version of statistical mechanics of an ideal gas. Here atoms and molecules of finite size contain the particle's properties and act as isolated islands of encapsulated primitive data with a volume, mass, charge, angular momentum, and magnetic moment while hiding other internal details.

2.2   Mathematics of a Physical Space

Our ignorance about quantum mechanics goes no further than our ignorance about the space we occupy. We are not living in space described by Euclid with a mathematical continuum. There are no points, edges, surface, and well defined regular polygons or simple parabolas defined by equations. These objects are all in the mathematician's imagination defined by point sets and the locations of the points are defined in a continuum. Infinite precision translates to an infinite energy requirement. Finite precision comes at a lower cost in terms of energy. We live in a fuzzy world where we don't have access to unlimited resolution. The world has to support data and matter with finite resources in terms of energy. The mathematical basis for finite resources is to limit the precision of any structure to support those ends, rather than supporting a set of precise regular Euclidean constructions. Classical Euclidean view of mathematics championed by classical physics quickly breaks down when trying to define a particle, because it is simply too costly a geometry in terms of energy to employ. For particles to have a finite set of properties the geometry used does not need to support infinite precision. A prime example is the precision of large G in the gravitation potential is only known to within four decimal places despite the experimental efforts to produce a more accurate result. Quantum measurements, being probabilities, defines our ability to acquire data about physical objects that are economically constructed. Limits to precision in laboratory frame measurements comes down to how the particle's own structure is represented.

2.2.1　Statistical Independence

There are a variety of theories swirling around trying to explain elementary particles at present. By adding a simple constraint to minimizing the excess energy in a theory independent of what is being described is the first step in producing a good model by reducing the overhead the theory requires to be functional. Trimming the fat not used by the equations to generate the theory by minimizing the data embedded in the geometry to model the physical space that supports both matter and radiation is essential.

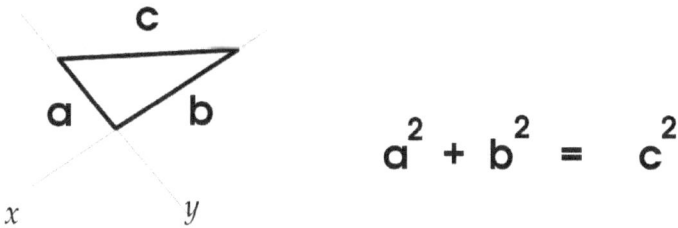

$$a^2 + b^2 = c^2$$

Figure 2.1: **Geometric version of the Pythagorean theorem that is like the quadratic relativistic relation for the conservation of energy.**

Above is a classic sketch of the Pythagorean theorem for two perpendicular line segments that lie in orthogonal directions. Disassembling the triangle and let a, b, and c float around and the only requirement is that a and b lie in some orthogonal direction in space that does not allow one to get a into b's dimension by a simple translation. The two dimensions that are perpendicular and orthogonal and at most can only share a single point. To sum orthogonal line segments lengths one must use the quadratic Pythagorean sum rule. It is not even necessary for the segments to have any points in common.

The Pythagorean theorem gets interesting when one realizes it is directly applied in the relativistic conservation of energy equation.

$$E^2 = p^2c^2 + (m_oc^2)^2 \qquad (2.1)$$

The equation is in terms of energy not length made up by the addition of two terms. The question becomes how can you add another term to the relation? Georg Cantor who in the 1870s explored continuum spaces answer would be to add another index to the index list of dimensions as there are no constraints on the number of dimensions for the continuum [49]. Not having detected these extra physical dimension it is apparent that one can not use the continuum. That did not stop physicists from adding a spatial dimension to the original three as this idea became a fad spawning innumerable theories with the adjectives of super and string attached. Fortunately, what works lies within the limitations of the three spatial dimensions.

In classical physics the conservation of energy was first determined by a well known experiment converting a defined amount of potential energy doing work into heat. When relativity came along the energy conservation law changed from being linear to being quadratic and looking very much like the Pythagorean theorem. That complicated the development of quantum mechanics by being a quadratic equation that almost everyone wanted to be linear. The terms in the quadratic equation are in units of energy that are made up of mass, scale, and time,

$$kilograms \ meters^2 / seconds^2$$

There is no restriction on the scale and time units being defined in the same space, for the kinetic and self-energy components, only that the energies should balance. The liberty of having separate space-time coordinates in different spaces is given by the expression itself. This freedom of expression was taken away when the space for quantum mechanics was assigned to a single space[1]. The equality of the conservation of energy relations is the connection between the spaces for kinetic energy and the self-energy.

Alternatively, there was another motivation as physicists like to have things simple, so they wanted a first order energy relation. To get a first order relativistic relation the square root of the quadratic conservation of energy relation is used [51]. Taking a square root was a grossly unnatural act because it alters the equation's functional form resulting in nonexistent negative energy particle. The conservation of energy equations are used as prototypes for generating the working differential equations of quantum mechanics. The relativistic conservation of energy is the prototype for a relativistic wave equation and the act of taking the square root changes its mathematical order from two to one with no physical justification. Dirac taking the square root eliminates the solutions that define both the fermion and boson fields. The square root is also not essential for the description of the anti-electron, positron, whose description depends only on the γ of special relativity taking on a negative value.

Pythagorean sum as it stands yields the conservation of energy for a particle in a potential free region. The importance of equation 2.1 is that the relativistic relationship is exact with mass being a real quantity and the other equations in Table 2.1 are approximations. The quadratic sum for the total energy is not as simple as it appears. It is a restrictive energy conservation statement where the mass is limited to being real for a particle.

Conservation of energy is a concept that needs to be carefully adhered to in its exact form and this is rarely done. There are at least four easy ways to violate the conservation of energy in quantum mechanics. The easiest is to use one of the two

---

[1]The assignment to a single space, then a single vector spaces was done by default with a series of contributors Bohr, Sommerfield, Klein, Dirac, Einstein, von Neumann, Wyle, and Foch.

non-relativistic expressions for energy as shown in the left two entries of Table 2.1. These are both approximations because they do not conform to relativity. The third way is to include a potential which is singular, $1/r$, having an infinite self-energy. The fourth is to tack on the self-energy to the Lagrangian.

Table 2.1: E is relativistic energy, **E** total classical energy, *T* is kinetic energy, *V* is potential energy, *L* is the Lagrangian energy, *p* is momentum, m_o is the rest mass and *c* is the speed of light.

Non-Relativistic Approximations	Relativistic Exact
$\mathbf{E} = \mathbf{T} + \mathbf{V}$ $\mathbf{L} = \mathbf{T} - \mathbf{V}$ ($\mathbf{E} \neq E$)	$E^2 = c^2 p^2 + (m_o c^2)^2$ $\downarrow \qquad\qquad \downarrow$ $E^2 = (Kinetic\ Energy)^2 + (Self - Energy)^2$

The potential energy, V, appears only in the left hand column for the non-relativistic cases, but not on the right hand side because not all potential energies can be treated in the same way.

The electrostatic potential can be defined by the partitioning of the self-energy as $V(r) = \delta m c^2$ where δm represents a small change in the particles mass. So that the relativistic conservation of energy relation can be expanded to include this potential energy where m_o is the rest mass.

$$E^2 = p^2c^2 + (mc^2)^2$$

$$E^2 = p^2c^2 + (m_o + \delta m)^2 c^4 \tag{2.2}$$

$$E^2 = p^2c^2 + (m_oc^2)^2 + 2Vm_oc^2 + V^2$$

From this conservation of energy equation that includes a self-energy dependent potential a relativistic wave equation can be constructed to replace not only the Schrödinger equation, but the Klein-Gordon and Dirac equations as well, section 4.4.

The relativistic conservation of energy is expressed in terms of the Pythagorean sum being applied to different kinds of energy: kinetic energy and the self-energy. The Pythagorean theorem is only a method of summing quantities that lie in orthogonal or independent spaces. A physical space where energy is defined requires independent spatial and time coordinates, otherwise using the same time variable would make the spaces linearly dependent. In each space the time-space coordinates are statistically independent lying in two different spaces with no algebraic connection. The key word is statistical meaning that any type of mapping by some set of equations like going from spherical to Cartesian coordinates would be totally scrambled. Statistical independence allows creating a dimension with the same physical space that is always orthogonal to the coordinates of the original space. The kinetic energy term of the laboratory frame is define in (R, θ, ϕ, t) or (X, Y, Z, t). The self-energy term is defined with coordinates (r, τ) where there is no algebraic connection between the two coordinate systems [16]. The self-energy space can take on spatial dimension three or less. It differs from the laboratory frame coordinates by being limited to spherical symmetry where the two angular coordinates are not available because they become randomized and being cyclical will not accumulate a value like the radial coordinate. That automatically fixes the symmetry of the self-reference frame to be spherical and described by the $U(1)$ group symmetry. It will be found that this truncated representation is a result of the two potential terms in equation 2.2.

Finding a statistically independent reference frame is the original rabbit out of the hat trick. The trick allows many copies of particles and fields to be made that are all mutually exclusive and statistically independent. It is the trick nature uses so that many particles can exist not just one. The Pythagorean theorem has turned into a more useful relation, rather than just being the end point for a course in geometry. The simplicity of the theorem is there is no other way to connect the properties in one space to the properties in another space because they are statistically orthogonal. To sum energies from the different spaces one is stuck using the Pythagorean theorem.

The self-reference frame's effect on the laboratory frame is to produce a distributed shrinkage so there is a volume deficit in the laboratory frame. Volume of the laboratory frame is the only thing of value that the laboratory frame can sacrifice. It appears there is a conservation law for space as well as for energy.

Gravity is a special global potential generated by a deficit in the volume of the laboratory frame. The sacrificed volume from the laboratory frame goes into building a particle/field self-reference frame. A deficit in volume also assures that there is no direct mapping that occurs between the spaces. Randomly defined deficit assures no analytic connection between spaces, because of the locally shrinkage of the laboratory frame that generates the gravitational potential. Though the self-reference frames of each particle are independent they are not totally isolated as they are coupled by gravity and the conservation of energy relation.

The orthogonality of spaces was missed in standard quantum mechanics by failing to notice that the space where the particle's self-energy, $m_o c^2$, is generated has to be orthogonal to where particle dynamics takes place. This mistake stalled the development of quantum mechanics at the end of the 1920s. The relativistic conservation of energy relation requires separate spaces and there is no way to get around this experimental fact for each independent particle because it defines the source of gravity between the macroscopic theory of general relativity and the quantum behavior of a massive particle. It also yields a relativistic quantum wave equation in the laboratory frame that now can support gravity.

2.2.2 Orthogonality

Like Gaul orthogonality is divided into three known parts. The geometrical form $x^2 + y^2 + z^2 = d^2$ where the sum of the squares of the three Cartesian components generates the distance from the origin, d.

Orthogonality's second form is exploited in the current description of quantum spaces, that is an orthogonality of functions, to create an infinite number of orthogonal function. This was originally applied by Fourier solving the problem of heat conduction. If the different functions are orthogonal then the integral of the product of different functions is zero. There are a number of different function pairs in addition to *Sine* and *Cosine* that share these characteristics. They all can create an infinite solution set for reconstructing arbitrary behavior. In the case of the hydrogen atom a similar function set defines the orthogonal orbitals of the atom through the spherical harmonic functions.

Orthogonality of spaces is what is required to understand matter and fields. These are statistically isolated spaces that have both a unique spatial and time dimension. The elementary problem for a photon is to understand the ability of fields to move through each other without interacting. This property is called superposition. Superposition would be easier to understand if each field was in its own independent space. The trick is accomplished not by being willed, but rather an active process is necessary to isolate the space of each field. This spatial orthogonality is an important concept that can be put to use in writing differential wave equations for both particles and fields in the space where the particles and fields are created, their self-reference frame (SRF).

2.2.3   The Inability to Conserve Energy

The standard model is an evolved collection of high energy theories depending heavily on the relation $L = T - V$ and the error of Pauli and Weisskopf in using the Klein-Gordon equation as a model for bosons [44].   The use of the Lagrangian was a non-relativistic starting point that voids the rest of their development. With a damaged model, any sense of reality is lost and unfortunately it was handed down to successive generations of theory students allowing more fanciful arguments to enter the subject.

The temptation to violate the conservation of energy must be an almost overpowering urge, because it was accomplished in so many different ways violating Wittgenstein's advice to say nothing without a good reason that dominated physics following WWII.

2.3   Physics is Conservative

Physics is conservative with a capitol "C" and not as carefree as a Cole Porter musical: conservation of energy, conservation of linear momentum, conservation of angular momentum, and conservation of charge. Symmetries are important, but they descend from the conservation laws and most particularly from relativity. Einstein took a long detour in 1911 and 1912 trying to show that light had the capacity to alter its velocity in a vacuum under the influence of external potentials. He came to realized this change generated non-physical results that conflicted with both special relativity and experiment.

The other item that seems to be conserved are real physical properties, while the Heisenberg uncertainty relations allow excursions in energy they don't allow a change in the net property content, eg. charge conservation, found at anyone time. There is no escape mechanisms for a new property such as charge to be randomly created as exists for energy. This conservation principle is violated routinely in the use of a single virtual photon to model high energy scattering from accelerator experiments as the connection between passing particles. The goal in this deception in tens of thousands of papers on high energy scattering was to avoid acknowledging a static potential as being a real structure.

The orthogonal spaces that separate the self-energy from the kinetic energy are useful in solving some simple problems that have been around for generations. If an object's properties can be encapsulated or generated where its self-energy is defined, this allows the primitive particle description to be treated as a simple object that can take on other properties such as a specific force field, linear, and angular momentum in the laboratory space.

The one physical property space has is its volume. Conserving volume is not a particular problem when you only have one space to occupy. When you embed multiple spaces into each other the question of volume conservation becomes important. This assumes your space has no embedded lattice, foam, spin, or chiral structure, as any of those attributes come with a change in the physical property content that is really an added energy content making them excess to a physical theory. The only property space can bargain with is its volume, and the manifestation of this is found in the curvature of the laboratory frame known as gravity.

2.4 Measurement

Any measurement requires the object to be measured and a measurement tool. For greater precision more expensive tools are required and a more precise standard. There is a cost in precision and it comes with the number of significant digits in the measurement, three place precision is much less expensive than five place precision. A general question about any measurement: Is there a limiting precision for measuring an object?

The answer is yes, there is a limit. The classical limit will be the precision you can measure the dimension of an individual particle in comparison to the smallest particle that can be used as a standard ruler. This standard scale will be represented by the symbol for epsilon, ϵ. If the particle is a point that cannot be measured that implies infinite precision is available at infinite cost. If the particle is not a point then it has a structure of which we have no description, but it will have a measurable dimension at rest of ϵ. In a single measurement the precision of the measurement will be approximately ϵ. That would be the minimum scale marking on a ruler if one could be made. This is a comparison with an identical particle to determine, ϵ. In multiple measurements a greater mean accuracy is approached by finding the mean of a set of measurements, but not in a direct single measurement. The final conclusion drawn from the measurements is that the real geometry of matter is not described by Euclidean geometry, but something less precise.

The limitation in measuring a particle go well beyond the analysis used to produce the famous uncertainty principle. The uncertainty principle does not come with a prohibition against $\Delta r \rightarrow 0$, the uncertainty of a scale measurement. A vanishing limit might not be possible in the physical world, where its always possible mathematically. Particles when compared to a standard possess a statistical character of its own, that will make a single scale measurement on the particle limited to a fixed threshold of uncertainty.

The question is whether this statistical imprecision is a fair game, a true random process. All games of chance we know about are compromised in one way or another. This is not a new problem as Marc Kac puzzled over this problem for a number of years [52]. Starting off by studying Brownian motion on a lattice that is called the random walk problem. That did not get him very far because the use of a lattice is a major handicap, as it automatically defines the space and the allowed action suffers from physically impossible constraints. Also a lattice introduces a specific symmetry and symmetries should be the end result of an analysis not imposed from the beginning.

He realized these problems with the lattice and moved on to study a quantum random walk with path integrals over many unconstrained paths as being less encumbered.

Mark Kac consider the concept of statistical independence of different physical variables as an important advance in his understanding of what can be measured [52].   He was motivated by a basic problem stopping physicists who were using high energy scattering experiments to determine particle properties and structure. From the failure of finding a relativistic path integral theory he realized that the entire experimental program had a serious limitation about what information could be extracted about the structure of interacting particles. There were limits to what could be learned from these experiments. To present this argument he wrote a famous paper *Can One Hear the Shape of a Drum* [31] where he showed there would always be an insufficient amount of experimental information to reproduce structure of the scattering target. This is also the default limiting problem for any AI program.

The questions that remained was: first, what is missing in special relativity that stopped their progress with the path integrals and second, how does statistical independence couple to spaces rather than variables? It is spaces that have to be understood that define the differential equations that describe the physics.

The statistical limitation in measurement overcomes the objection of A. Einstein's about the statistical basis of quantum mechanics and gods playing with dice. The dice are fuzzy. A particle's individual limiting statistical uncertainty in location cannot be forced to go to zero because it is an impossible measurement. In contrast with Newtonian mechanics or general relativity, which allows one to compute exact orbits, this single particle fuzziness cannot be ignored. The orbits may be exact, however, the precision for measuring the particles in those orbits is still limited by a standard of comparing two identical particles. Einstein was correct about quantum mechanics being incomplete [4], but not about its statistical basis being flawed, statistical independence will be the solution to the problem of how properties and the particles themselves are created and preserved in the spaces we live.

Feynman's approach to quantum electrodynamics was not much different than Einstein's approach to general relativity in that they both used a space that would have been familiar to Euclid, though warped. They both adopted the mathematical continuum and not a physical space for their models. The laziness of applying a known geometry instead of working out the details of space from energy conservation was the principal mistake made by physicists of the 20th century. Unfortunately, the education of these 20th century theoretical physicists made them good students of Euclid. The experimentalist had the advantage as it appears the fellows doing quantum mechanics had forgotten the words of Riemann and Gauss that the properties of a physical space needs to be discovered by experiment.

The question is why there is a limiting precision to measuring a particle's dimension like the electron? We are going to give our experimental students help with the algebra that is not too involved and was available in 1905 [16] [35]. The argument is very simple if you take the empirically tested kinetic version of Einstein's quadratic relativistic conservation of energy using m_0 as the rest mass, p the momentum, c the speed of light, v the particle velocity and start with the square of the energy, E found in equation 2.2.

$$E^2 = m^2 c^4 = \gamma^2 m_0^2 c^4$$

$$E^2 = \frac{1}{1 - \frac{v^2}{c^2}} m_0^2 c^4$$

$$E^2 (1 - \frac{v^2}{c^2}) = (m_0 c^2)^2 \qquad (2.3)$$

$$E^2 = \frac{m^2 c^4 v^2}{c^2} + (m_0 c^2)^2$$

$$E^2 = (pc)^2 + (m_0 c^2)^2$$

What is interesting about this simple energy relation is that it was ignored until the middle of the 1920s when O. Klein and W. Gordon put the Klein-Gordon equation together and when Dirac worked on his relativistic equation for quantum mechanics. Even Erwin Schrödinger toyed with the relation [51]. All parties made a mess of its application. Einstein appears to have completely ignored this relation probably on purpose as it implied a general nonlinear behavior in energy conservation.

If you check a modern text [53] [54] [55] the quadratic relation for the relativistic energy is derived from the four vector relations $(p_x, p_y, p_z, E/c)$ that was modeled on the work of Minkowski on a space for special relativity where he introduced the 4-space: (x, y, z, ict). There is a problem with the vector containing both momentum and energy. It implicitly assumes there is a single time variable that will be shared by all the components of this 4-vector. The self-energy of a particle does not involve the kinetic behavior of a particle. The quadratic expression for the relativistic conservation of energy, implies the self-energy possesses an orthogonality to the kinetic energy. The orthogonality then also holds for the coordinates that make up the self-energy. Meaning the time and space coordinates are different for the kinetic energy and the self-energy terms. Even though the rest energy possesses units of time and space in its make up, it is not the time nor space of the four vector (x, y, z, t) and the rest energy has no relation to momentum at all that is found in the four vector $(p_x, p_y, p_z, E/c)$. $(p_x, p_y, p_z, E/c)$ is not a true vector in the laboratory frame but a composite structure used for convenience. The quadratic derivation from equation 2.1 is correct, whereas deriving the same quadratic relation from the norm of $(p_x, p_y, p_z, E/c)$ is not valid. What is being hidden in this modern derivation are the spatial and time coordinates that make up the self-energy as they are independent of laboratory frame coordinates.

What this directly implies is that the particle property of self-energy, $m_0 c^2$, is derived in an isolated independent space apart from the laboratory frame. However, this isolated space does not have a physical boundary to separate it from the laboratory frame rather its coordinates are statistically independent from the laboratory frame so the particle is not hidden. Our test particle has a limited measurable resolution of epsilon, ϵ, in the laboratory frame because it cannot be exactly mapped into the laboratory frame. This means that our test particle represents a truly random process tracing its center of symmetry displaced by ϵ on any successive measurement when it is regenerated even at rest. The algebra of this fuzziness is included automatically in the relativistic energy conservation relationship when reduced to a wave equation. The fuzziness is part of the random process's ability to regenerate a particle and its properties. Our space, the laboratory frame, is not the perfect space of Euclid, it is an imperfect space that it shares with statistical independent spaces, so particles can have a set of coordinates to generated their own structure, not connected by an exact mapping to our own space, the laboratory frame, where measurements are performed. The key feature is a boundary between spaces that is statistical in nature, where no new material has been used to fabricate the boundary. This form of boundary does not require any of the special mathematical properties required for defining a domain in the continuum. Statistical independence does not generate a closed domain only a functionally separate domain. This minimalist approach to defining physical properties allows for multiple particles and fields. All of this can be formally presented in a set of differential wave equations for both spaces that are coupled through a random process of particle creation and annihilation [17].

Physicists have worried about hidden variables in quantum mechanics for years in trying to reconcile relativity with quantum mechanics. The most famous attempts by J. S. Bell with his inequalities proved particles and fields do not behave as if they have a specific locality [56]. Experimentally, this was no surprise. The physical example of this is the Young double slit experiment where a particle or fields appears to be in two places at once [57]. What they did not consider was having hidden spaces in plain sight. Statistical independence generates an encapsulating mechanism different in function than a cell wall even though both allow information about what is captured in that isolated spaces to produce physical properties. The cell boundary material is added inventory for making a cell. There is no added inventory in forming a boundary to produce a statistically independent space to house a particle. There is only a local reduction of the free volume of the laboratory frame that is supplied to the self-reference frame. The only property that the laboratory space possess is its volume and that is all that can be sacrificed when there is matter about.

Traditionally, in physics one either derives or discovers a differential equation. In this case it was started by including the random process of pair-production operating on a massless longitudinal field [46]. The derivation was triggered by an experiment that was able to observer activity within the self-reference frame of the particle [23] [16]. After solving these equations slowly one starts to understand where the solutions lead. This is typically how the discovery process works when doing physics. A description of matter and fields are all generated at the same time from a statistical basis. What was not planned was finding an independent space, (r, τ), which is different than our three dimensional space, (X, Y, Z, t). This restriction on accessible coordinates in the self-reference frame immediately solved the paradox of the Young double slit diffraction experiment.

The calculations show that a trade off is made for producing particle properties in our physical space by sacrificing the ability to use some coordinate variables in the particle's space and some of the volume from the laboratory frame. These new spaces make a strong symmetry statement in that whatever occupies their individual frame of reference has a spherical symmetry. The new space that defines the particle's scale simultaneously limits the precision in measurement. The laboratory frame gains gravity as a measure of the particles it contains.

2.5   The Second Quadratic Equation

When the self-energy is expanded into a potential in the conservation of energy equation there is a second quadratic relation that appears with the potential, V. This relation is essential in making the connection between the particle's self-reference frame and the laboratory frame found in Table 2.2.

The connection between the two spaces is via the potential terms. Setting the potential terms equal to zero generates two solution that introduces the mechanisms of pair-production. Pair-production is a property conserving process randomizing not only the location, but also renewing the particle. Whether the initial particle or its pair produced twin is annihilated generates the basis for the random process.

The process leaves no residuals, only a particle with an uncertainty in its displacement equal to its scale. When played out, the connection between the laboratory frame and the self-reference frame generates the scale of the particle, ϵ in the form of the Compton relation, $\epsilon = \hbar/m_0 c$, in the laboratory frame [17] [58]. The action of renewing the particle does not occur over all space but only where there are particles that get continually renewed endowing the particle with inertial mass. There is no

Table 2.2:   **Randomizing process of pair-production initiated in the laboratory frame and carried out in the self-reference frame to renew a particle while generating its inertial mass [17] [58].**

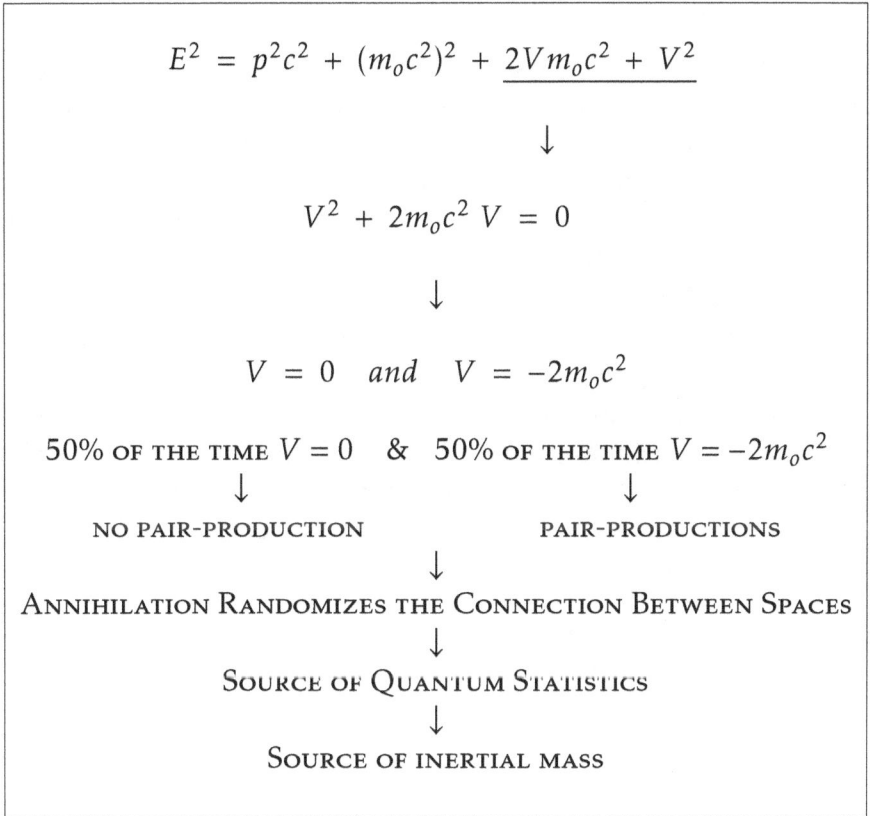

$$E^2 = p^2c^2 + (m_oc^2)^2 + \underline{2Vm_oc^2 + V^2}$$

$$\downarrow$$

$$V^2 + 2m_oc^2\,V = 0$$

$$\downarrow$$

$$V = 0 \quad and \quad V = -2m_oc^2$$

50% OF THE TIME $V = 0$ &   50% OF THE TIME $V = -2m_oc^2$

$\downarrow$ $\downarrow$

NO PAIR-PRODUCTION   PAIR-PRODUCTIONS

$\downarrow$

ANNIHILATION RANDOMIZES THE CONNECTION BETWEEN SPACES

$\downarrow$

SOURCE OF QUANTUM STATISTICS

$\downarrow$

SOURCE OF INERTIAL MASS

seething sea of particles coming in and out existence in the vacuum, there must be a seed, a particle or a field. That implies there is no independent vacuum energy. This renewal process removes access to the angular coordinates of the self-reference frame because they are cyclical coordinates and will not accumulate a measurable value. The resultant spherical representation in the self-reference frame fixes the spherical symmetry of the particle as the principal symmetry.

2.5.1 Gravitational Mass

A long standing problem was the question of why gravitational mass should be identical to inertial mass. The reason for their equivalence appears to be in the dynamic origin of particle's self-renewal and the volume of space required for the process where both inertia and the gravitational potential are generated. Simultaneously, the gravitational potential, V_g, is a product of the space acquired from the laboratory frame and will have to be treated differently in the differential equations of quantum mechanics so as not to interfere with the process that generates inertial mass.

The gravitation potential, V_g, is expressed by general relativity, a macroscopic theory, as a curvature of the laboratory frame, whereas, the electrostatic and strong force potentials have their origin in the structure of a particle. This requires that the gravitational potential a particle has acquired directly alters the total energy. The conservation of energy expression is different than that which includes the electrostatic potential.

$$(E + V_g)^2 = p^2c^2 + (mc^2)^2$$

$$(2.4)$$

$$E^2 + 2EV_g + V_g^2 = p^2c^2 + (mc^2)^2$$

2.6 Why is Space Uniform

If space is not isotropic angular momentum is no longer conserved. Action at a distance was a problem taken up in Maxwell's major work *A Treatise on Electricity and Magnetism* [29]. The problem was defining the spaces occupied by particles and fields supporting both dynamic and static action. The old agent, the aether, was something proposed to reconcile the conflict in producing a force at a distance. The list of those exploring this conflict included C.F. Gauss, W.E. Weber, W. von Helmholtz, W. Thomson, B. Reimann, C. Newmann, R. Clausius, E. Betti,

R. Cotes, and finally E. Torricelli on the ether whom Maxwell quotes:

"is a quintessence of so subtle a nature that it cannot be contained in any vessel except the inmost substance of material things" Lezioni Accademiche (Firenze 1715) p. 25

This early sentiment contains a major truth. Maxwell was stuck with the point charge and its potential without enough time to explore Torricelli's idea, but he was looking in the right direction. The point mass and charge had to be connected to the two different field types, static and dynamic. The inability to reconcile the properties of the static potential with dynamic description was his problem. He had no trouble with dynamic fields, it was the static ones that were the problem. Point masses and point charges require some magic to produce an action at a distance in generating a potential. The concept of an aether filled that void until a structure could be found to support the static fields. The structure is found in the particle's self-reference frame state function where for the electron its static electric field is computed to be its own normalized structure in space. The electron's electrostatic field is the major contributor to the structure of the electron that is now no longer a mathematical point.

The private space for a particle would have to be spherically symmetric, loosing the connection to the angular coordinates. The electrostatic field of the electron is spherically symmetric [30], meaning if the particle suffers any rotation about its center of symmetry there is no apparent change in the field structure. This feature maintains the spherical charge symmetry forcing a spatial dependence for the particle's structure dependent only on the radial distance r from the particle's center of symmetry. The electron would not be expected to have the same structure in its private space as the photon, because the electron, as a longitudinal field, generates a radially oriented electrostatic field that is different than the photon's transverse field. The static field of a particle now can be defined by its structure from its self-reference frame solution [35].

There is an important point of mathematics that goes back to Cantor and his study of the continuum [49]. Dimensions are assignable with any indexing scheme for the continuum. In the continuum dimensions are arbitrary, 2, 10, or 100, it makes no difference it is only a matter of selecting an index. This has been taken advantage of since Kaluza and Klein tried to add a fifth dimension to the four dimensions to weld electromagnetism into general relativity. The pattern has continued with the differing forms of string theory adding dimensions. The existence of the self-reference frame for individual particles and fields automatically excludes the use of the continuum as a valid description of space and its arbitrary definition of dimensions. The separate self-reference frame entities are all visible in the laboratory frame with some limitations on precision. The existence of different self-reference frames embedded in the laboratory frame puts to rest the many worlds interpretation of quantum mechanics [59]. Nature found a version that actually works where the volume of space appears to be conserved.

Spherical symmetry can be linked to the isotropy or the sameness of space every direction. Isotropy of space is a feature displayed by the behavior of particles and fields that occupy an empty space. The mathematical spaces one uses for quantum mechanics to describe particles and fields operates on the complex plane. The use of the complex plane is a natural result of solutions of the second order differential equations produced by the quadratic conservation of energy both in the self-reference frame and the laboratory frame. Complex wave functions are what is required to represent the different types of longitudinal and transverse fields if one wants to understand matter.

The Russell/Whitehead program of applying mathematical rigor to a field failed with arithmetic. Physics did not function well with a set of axioms authored by von Neumann. The mathematics and logic of Kurt Gödel's recursive programming seems to underlie the basic engine of sustaining matter. Confusion about quantum mechanics was not settled by applying the few partially working equations, because they were not correctly derived. It took an experimental examination of large scale quantum fields that exposed the self-reference frame, which settled the problem of realistic spaces that can support massless fields, inertia, and physical properties.

2.7   Where Organizations Fail

The errors made in basic quantum mechanics in the 1920s produced debates which persisted into the early 1930s, but were insufficient to correct the defects. One problem was the principals working on quantum mechanics did not take a wide view to bring in other relevant field data that was available. The second and a more persistent problem was that physics was now being organized as an essential enterprise.

In the first decades of the 20th century colleges started hiring professional managers and started calling themselves universities. The brakes were applied to the free spirit of investigation of the late 19th century, right when new research areas started taking off, so these new administrators could be fed. The economics of the university system proved a formidable hurdle for anything revolutionary, unless it could produce income for these new academic managers. Later in the 20th century ideological constraints were imposed on physics by narrowing the political range of allowed faculty appointments [1]. The sciences suffered in this regime. Finding errors or changing one's mind about the basics was now not looked on kindly by this new administrative class of managers whose reputations would be questioned. Hence, the Orwellian dictate from journal editors of having no citations older than 5 years for physics papers was the solution to maintain the status quo. Aberrant behavior of dealing honestly with data would make life too difficult for the ideological and financial parasites operating the universities.

Physics cannot be shown to be a sham as the public would not soon forget the folly of these managers. It became necessary to insure these managers would not be caught out by changing their minds. Here CERN and the DoE national labs played a major role of setting an agenda that could not be experimentally challenged by supplying editors to all the major physics publications.

The universities focus on raising research income made teaching a secondary product and this eventually damaged the ability of the public to challenge the academy. The transition occurred over a few of decades from the late 1940s to 1970. P. Kusch tried to push back against this by pointing out university research should only take up critical problems to physics and not be caught up in the minutia of being paper mills. However, university administrators were supported by the paper mills that ensured no controversy would be of any significance. With the rise of the DoE national labs, IASs, and CERN, real research was decimated along with educational significance of the research universities. Their misinformation was soon passed down to colleges, secondary and eventually primary educations. Now we have a system where administrators out number students and real research has gone back to were it began, to the independent amateurs.

Questions

1) New structures found in nature can be called emerging structures, because they encapsulate a set of properties unique from their underlying material. How can this process be described? Give some different examples?

2) During an elementary scattering event with a head on collision is it possible to produce a state of tension? Why is this question being asked?

3) If you were asked to design an unstable particle what characteristic would the particle be given? Think of a proton being crushed.

4) Based on what is presented here is there an alternative to the

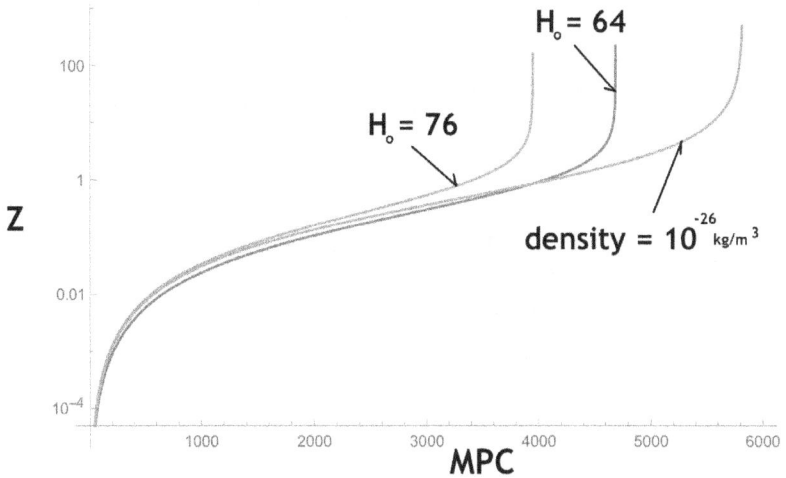

Figure 2.2:   **The mean density of 10^{-26} kg/m^3 computed for the gravitational red shift falls between the two extreme measurements of the Hubble constant [60].**

theory to explain the red shift that has been commonly thought of as a product of the universe expanding?

5)   The self-reference frame and the laboratory frame looks much like two parts of a recursive loop found at the center of all operating systems. The question is why would this be the minimal structure to define the existence of a particle? What are its advantages?

6)   Secondly, how would this active loop of a particle explain how the particle adjusts to the presence of an external potential?

Chapter 3

Consistency and Debris

"THE GEOMETRY OF PHYSICAL SPACE NEED NOT BE GOD-GIVEN EU-
CLIDEAN SPACE, BUT COULD WELL BE SOME OTHER GEOMETRY AND
SHOULD BE DETERMINED BY EXPERIMENT, NOT BY HYPOTHESIS"

BERNHARD RIEMANN [61]

Quantum mechanics stalled at the end of 1920s because the relativistic conservation of energy from A. Einstein's special relativity was viewed as something strange and the questions it raised were not engaged. It was still a world steeped in the linear conservation relations of thermodynamics. The key mistake made by Dirac, which the high energy modelers took advantage of was to ignore the self-energy in the quadratic conservation of energy relation and assume that for matter like light the energy, E, is proportional to p, particles momentum, $E \sim p$, see Figure 3.1 [34]. It was exactly the same starting point for our work, except the field studied was not transverse as is light but longitudinal.

The end result produced the self-reference frame for massive particles and their state equations that describes the massive particles' structure that results in their inertia. The self-energy $m_0 c^2$ is the end product of the derivation that starts with a massless longitudinal field. We had the advantage of having an experimental window on the self-reference frame via the longitudinal spin wave [23]. The quadratic relativistic energy directly produces the relativistic wave equations that function in the laboratory frame where things are measured. Something the Schrödinger equation fails at and the Dirac equation tries to force.

General relativity somehow gave license to bypass the problems that special relativity presented. In our search through

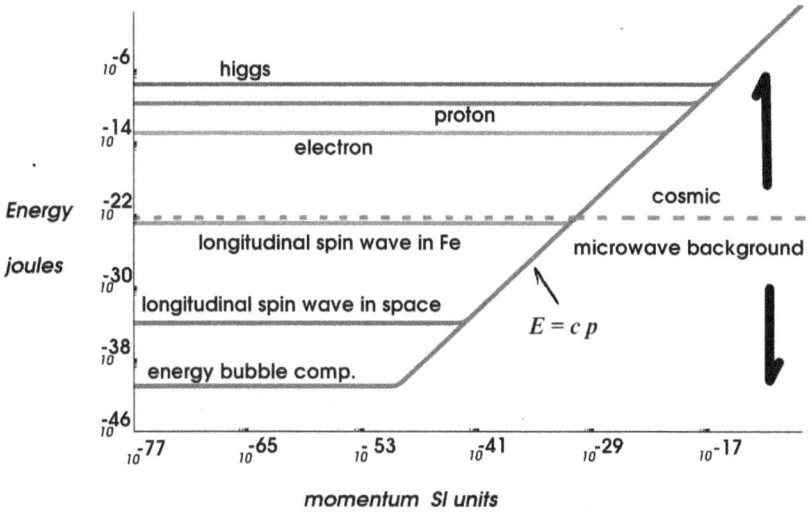

Figure 3.1:   **After Dirac's mistake of adopting a first order wave equation from, $E \sim cp$, all high energy physics operates on the sloping line to the right while ignoring the particles low energies origins to the left. The cosmic microwave background appears to be a dividing line between intrinsically massive particles and collective excitations that have inertial mass. [22]**

the literature the only person to generate a relevant comment about special relativity was John Stewart Bell in a BBC interview [62]. He had realized there was a problem with the way special relativity had been handled. Dirac's false start to quantum field theory using a "$E \sim p$" for massive particles with the wrong polarization was the oversight, thus releasing theorists to populate the world with models of physics and cosmology that had at best a tenuous connections to reality. The standards for examining experimental data, especially high energy scattering data were often ignored [14]. Financially, it was more profitable to defend an idea that could not be experimentally tested and this route became a favorite of both researchers and funding authorities. In both cases no one would be wrong and they were not spending their own money [63].

3.1   Too Little Experimental Data

In the 1920s and 1930s computing the hydrogen atom's ground
state energy was the validation for both the Schrödinger and
Dirac equations. Polykarp Kusch's strategy to attack the rise
of quantum electrodynamics in 1966 was to use the failure of
those calculations to accurately explain the ground state energy of hydrogen. There was a large spectroscopic discrepancy
in both the Schrödinger and Dirac energies for the hydrogen
ground state as shown in Table 3.1. When this discrepancy
is explored for ions with a single electron and greater nuclear
charge the models of Dirac and Schrödinger break down in a
more dramatic fashion [13].

Table 3.1:   **Hydrogen ground state energy eV. The γ is
from special relativity and for a free particle is equal to
$1/\sqrt{1 - v^2/c^2}$, however, for a bound state particle γ takes on
values less than one meaning the planetary model no longer
holds for atoms. See section 4.1.1 for relativistic derivation.**

Schrödinger 1926 [64]	Dirac 1928 [3]	Relativistic 2024 [13]	Exp. 1971 [65]
-13.5983	-13.5985	-13.5956	
dev -.0033	dev -.0035	dev. -.0006	-13.595 eV
$\gamma = 1$	$\gamma = 1$	$\gamma = .9997339$	

Not until the correct spaces are used, self-reference frame
for particles and fields and the laboratory frame for dynamics
and measurement, along with the notion that special relativity
was complete was it possible to compute an accurate ground
state without applying the non-unique tools of quantum electrodynamics.

3.1.1   Inertia & Special Relativity

Newton's **F** = **m a** does not look like a relativistic equation, but looks are deceiving. The great question about mass is: where does physical inertia come from? That can only be answered by taking a look of how relativity plays a role in defining the basic mechanism that sustains and renews stable particles. Massive particles are made up of longitudinal fields whose wave functions are generated in their individual self-reference frame and they are sustained in a continual renewal process using pair-production in the laboratory frame [22]. Any stable entity is really only the manifestation of an on going dynamic process between the spaces where the entity is defined, see Figure 3.2.

3.1.2   Origin of Particles and Fields

From the relativistic massive particle description in the self-reference frame with a longitudinal field described by second order differential equations, there are two solutions: the fermion and the boson. The detailed solutions are found in section 4.4.2. Spin is not a property of the self-reference frame, but a property generated in the dynamic laboratory frame, section 4.1.2. From the standard solutions of a particle and field in the laboratory frame the spin angular momentum for stable particles can be computed [17]. From the two solutions for massive particles from the wave equation in the self-reference frame a transformation can be made to derive the transverse massless fields of both the massless bosons and fermions.

The four 3D solutions from the self-reference frame describe four elementary particles and fields: the electron/positron, the electron neutrino, the photon, and the massive boson, $W^{\pm}Z^{o}$, which has the ability to change the sign of its weak charge [16]. These four three dimensional solutions to the quantum equations are products of the self-reference frame. There are also solutions for one and two dimensions in the self-reference frame that must be considered. Assembling the one and two dimensional components will produce the other two neutrinos, muon, tau, neutron, proton, the high energy baryon states, and

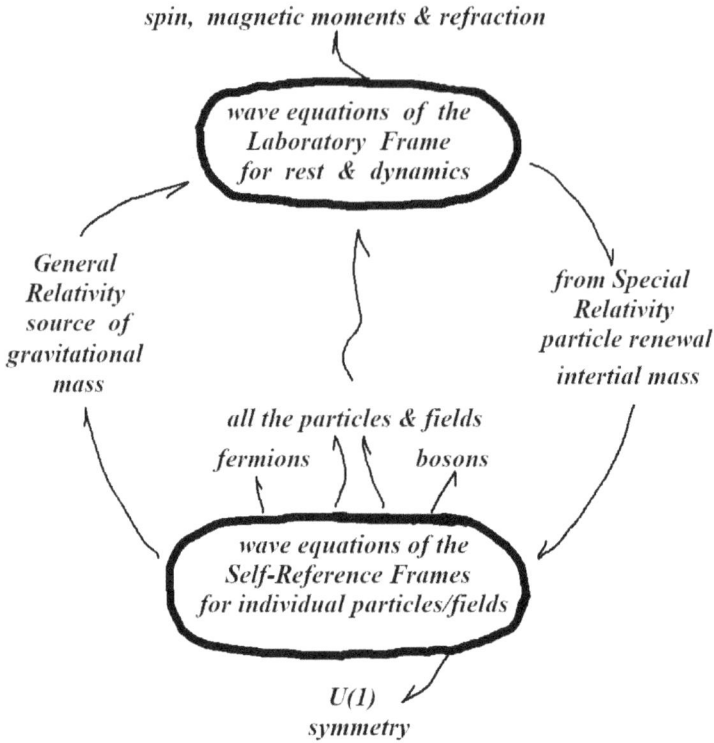

Figure 3.2: **There is a continual connection between the particle's self-reference frame and their presence in the laboratory frame through their dynamic renewal process.**

the mesons [22]. The assembly of the composite particles occurs without the use of gluons [66].

Fermion fractional charge quantization comes directly from the lower dimensional equations in the self-reference frame [16]. The lower dimensional components do not appear individually but in combination to make up three dimensional particles. The common elementary particles are generated by correctly dealing with quantum mechanics should be of no surprise as it was predicted by A. Einstein [47].

The boson/fermion designation for particles/fields are generated from the 3D solutions in the self-reference frame are very specific. The assembly of one and two dimensional components into the the mesons, baryons, and leptons use a mixture of both fermion and boson components. The restriction spin imposes on allowed states for stable boson and fermion fields is an associate property of the laboratory frame alone. Spin strictly characterizes allowed stable states limited by the few allowed derived angular momentum states for particles and fields, see Section 4.1.2. Boson and fermion properties are fully expressed in the bound state via the Pauli exclusion principal for fermions and Bose-Einstein condensate for massive bosons, see Figure 3.3.

3.1.3   Charge Quantization

For massive particles, charge quantization is a property generated in the self-reference frame by examining the phase dependence of particle's complex wave function dependence on the relative energy, γ. For the fermion the phase dependence generates a quantized set of charges that only depend on the number of spatial dimensions. The allowed charges derived are 1/3, 2/3, and 1, for dimensions 1, 2, and 3 respectively. The boson charge is a continuous variable that is not quantized producing the characteristic CP violation, charge-parity, symmetry failure of the weak charge [16]. Spin and the magnetic moment are generated in the laboratory from the rotation of charge density. The fundamental field in three dimension for the electrostatic field is made from the product, $\mathbf{E} \sim \mathbf{u}^*(r, \kappa)\mathbf{u}(r, \kappa)$ of the fermion wave function in the self-reference frame for the three dimensional electron. For the composite baryon, mesons, and leptons with mass it is the product of the massive and massless boson components that generate the electrostatic field [22].

3.1.4   Symmetry, Time and Anti-particles

The density function has a spherical symmetry, $U(1)$, for all solutions in the self-reference frame where only the radial coordinate is accessible. The spatial coordinate, r, is coupled with

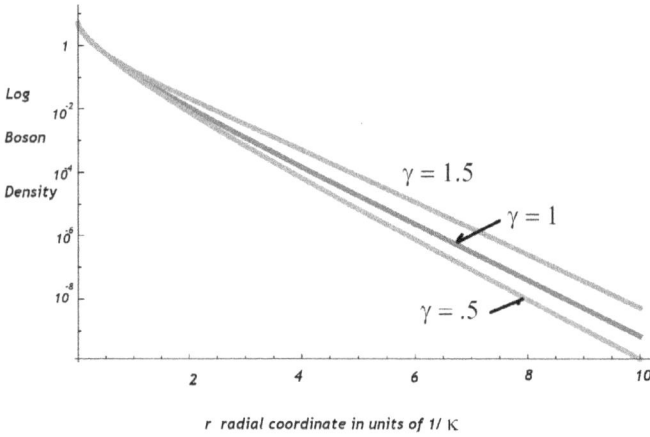

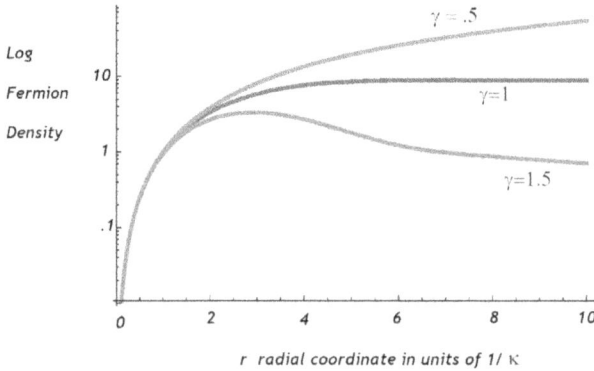

Figure 3.3: **Origin of fermion repulsion and the boson attraction that underpins the Pauli principle and Bose-Einstein condensation shows that fermion state function expands when the relativistic γ is less than one in the bound state and just the opposite for bosons in a bound state.** [13]

a private time variable for each particle/field in its own frame of reference moving to more positive values as each clock starts with the particle's creation. The time variable in the laboratory frame is under no such restriction. The anti-particles are nothing more than particles having an opposite handedness of its track on the complex plane of their self-reference frame that satisfy the equation $E^2 = \gamma^2 m_0^2 c^4$ that can have a negative value for γ. The anti-particles are not negative energy entities and are not traveling backwards in time. In the self-reference frame the γ of special relativity can take on negative values that generate the representation of massive anti-particle and also can have values less than one in a bound state, $|\gamma| < 1$. A plot of the the 3D fermion solution of the wave equation, $\mathbf{u}(r)$, in the self-reference frame for massive particles: electron and positron are shown in Figure 3.10

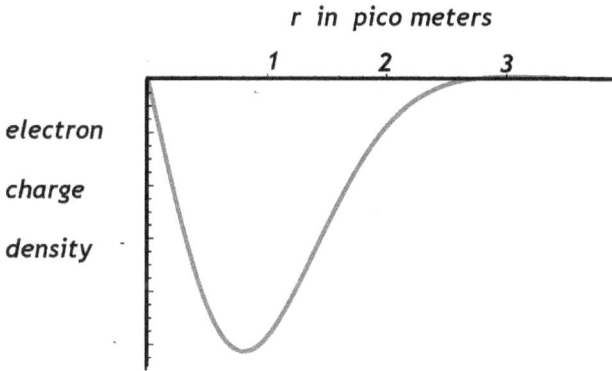

Figure 3.4:   **The charge distribution of the electron is computed from its electric field $\sim \mathbf{u}^*(r)\mathbf{u}(r)$ by Gauss's law [35].**

Spin

Spin is no real mystery once the point particle concept is abandoned where spin that is activated in the laboratory and reduces the self-reference frame's U(1) spherical representation. It is simply the rotation of the field structure in the laboratory frame moving at the speed of light that reduces the particle symmetry, meaning its symmetry is a secondary derived property and is not fundamentally stitched into the space that gave rise to the particle.

Once the structure of either a massless field or a massive particle in three dimensions is known it is possible to compute

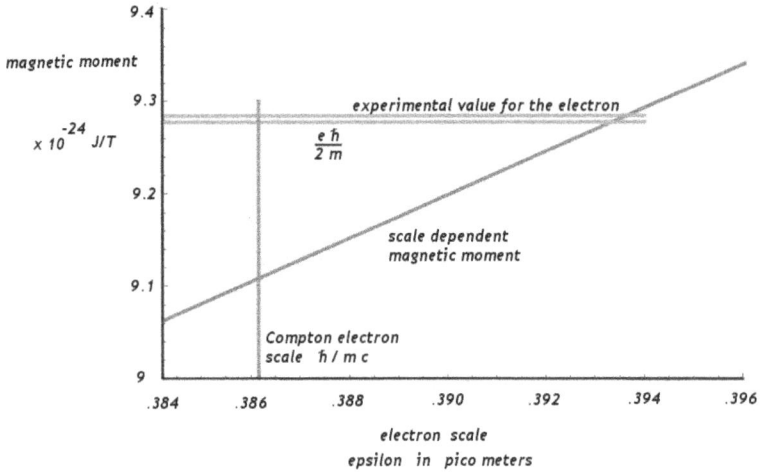

Figure 3.5:   **The vertical line on the left is the Compton scale from the electron's mass. However, the electron scale size required to generate the magnetic moment is greater by $\sim 1.89\%$. This is also a measure of the fraction of self-energy partitioned to both the angular momentum and the static magnetic field of the electron that is significantly greater than the self-energy in the electrostatic field which is only $\sim .13\%$. [22]**

the angular momentum. In the case of the photon when its lab frame wave function is integrated over its wave front the angular momentum comes out as $\hbar$. The neutrino has one half the density of the photon and therefore its angular momentum is $\hbar/2$. The three dimensional massive fermion either the electron or the positron when integrated over its rotating structure produces $\hbar/2$ for its angular momentum. In all cases there is no special requirement made on the structure of spaces for particles and fields to support an intrinsic angular momentum. Angular momentum is a dynamic property computed in the laboratory frame and see section 4.1.2 for its derivation [17].

3.1.5 Baryon and Mesons

The 1 and 2 dimensional components of fermion and boson fields go into the construction of the baryons, mesons, muon, tau, and their neutrinos. There are a total of 8 different components in one and two dimensions. These components are not quarks or gluons and possess no color flavors, but the solutions for both the massive and massless fields in their self-reference frames. These components are capable of constructing both the electrostatic fields and the strong field components that interact via contact. There is no requirement for gauge bosons or axions to describe any of the forces [66].

One interesting result is the charge neutrality of the neutron is independent of the scale variations in the set of components that generates its electrostatic field, where all the components share the same center of symmetry [17]. The neutrons charge distribution is also adequate to generate its magnetic moment. Another early deduction was that the τ particle is not a charged lepton, but neutral that is consistent with the original experiment at SLAC that found an excess of pair production products, the distinguishing feature of neutral particle decay, at the τ production threshold energy[22].

3.1.6 Dimension Zero

Dimension $n = 0$ becomes important for collective states that have to treated as a whole. A stable state will have its wave function decreasing with increasing, r, see Figure 3.6. That stability is found only experimentally for the Bose-Einstein condensation massive boson allowing superconductivity [67] and ferromagnetic longitudinal spin waves [23].

3.1.7 Nuclear Forces

The mechanism holding deuterium together is equally shared by magnetic dipole binding with the strong force. The nuclear energy binding curve has two regions: one for the light isotopes where the binding energy per nucleon increases with

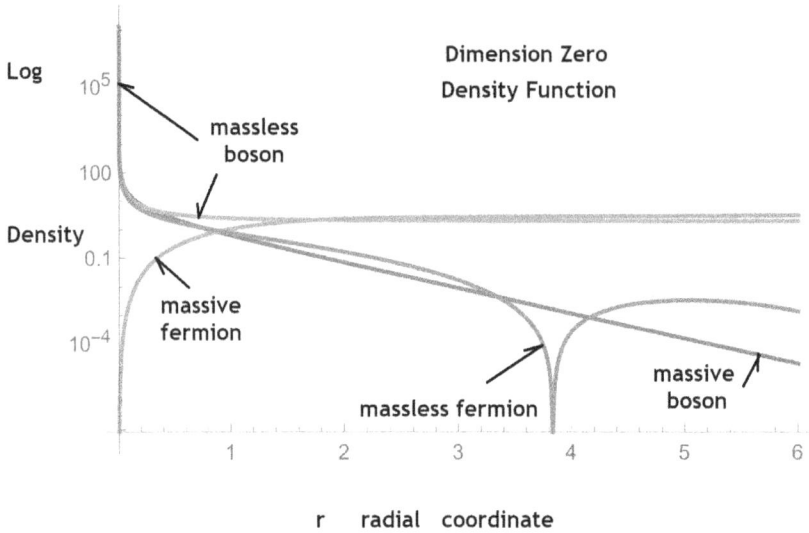

Figure 3.6: **The massive boson density function in the self-reference frame for dimension zero is the strongest species to form a collective state and the only one experimentally shown to exist[10].**

increasing atom number up to about neon and then begins a slow decrease for the heavier isotopes. The lighter nuclei behave very much like planetary type binding where it appears in some cases three body chaos plays a role supplying instability. The binding for the high Z nuclei is very different taking on characteristics of a condensed phase.

The protons have an electrostatic field that appears to be sourced by both the massive and massless bosons components of one and two dimensions [22]. Where the strong force is produced by the one and two dimensional massive fermion components, see Figure 3.7.

A case can be made for both a nuclear electrostatic force and the strong force being a contact interactions based on the field structure of their boson and fermion components for a prompt interaction without the requirement of a shuttling gauge boson. The limited inventory of lower dimensional components and

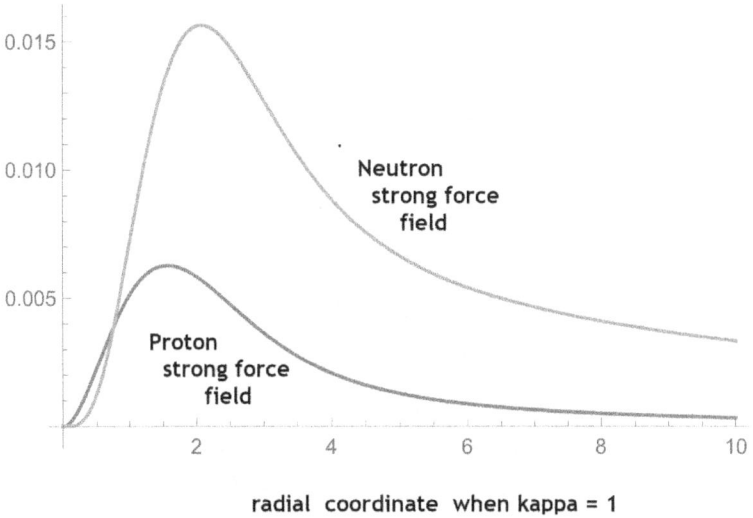

Figure 3.7:   **Strong force field for γ = .88, for the neutron and proton. The fields for the particle and anti-particle must have opposite signs and they are attractive. From the fact that neither proton-proton or neutron-neutron pairs are commonly found it is expected the the proton and neutron fields have an opposite sign and are attractive to each other reducing their energy when bound, taken from [66].**

the forces they build eliminates the magnetic monopole and the axion as possible particle. The nuclear electric field seems to be fundamentally defined by structure with spin generating the magnetic moment, and there is no CP problem with the way nucleons are bound [66].

Weak Force

The weak force has more in common with a pair-production process where the charge is altered to preserve the long range electrostatic field. The weak force is analogous to a phase transformation between two states where the structure is altered in the transformation [68]. The weak force is most commonly characterized by the decay of the neutron where there is a pair production of oppositely charge particles of different mass leav-

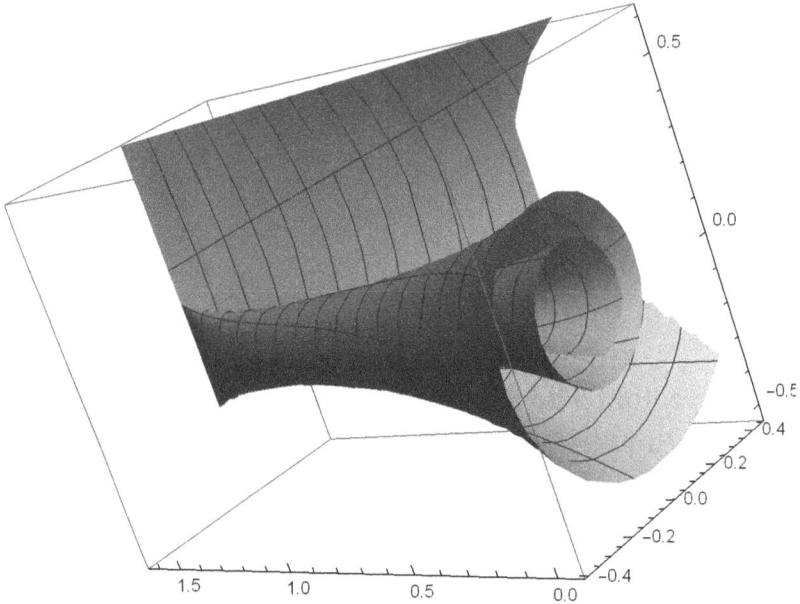

Figure 3.8: **The spiral on the front face is the electron's trace on the complex plane for its representation u(r) in the self-reference frame. The straight line on the rear face is that of the electron neutrino, ν_e, in its self-reference frame. The transformation that takes the electron into its neutrino is varying θ from zero to $\pi/4$ in the transformation of mass, $m \to me^{i\theta}$ [16]. The $e^- - \nu_e$ ensemble would be a massive spin zero boson inheriting the CP problem of all massive bosons when operating to generate a weak transition.**

ing a residue of an electron neutrino, v_e. Because the event takes on the order of 17 minutes to occur it is very unlike most prompt nuclear reactions that occur in less than one microsecond.

The weak polarization of the multicomponent nuclear matter [45] may give rise to an electron-electron neutrino pair polarization, $e^- + v_e$, as the agent generating the charge transfer. The electron and its neutrino are coupled by the transformation that takes a longitudinal field into a transverse field that can only be completed if there is a change in the charge to balance the charge change of the nucleon. The concept of a massive high energy $\sim 80 GeV$ $W^{\pm}Z^0$ being involved in this particular weak decay seems unrealistic.

3.1.8 Superconductivity

Figure 3.9: **Data taken from 1 kHz to 1.5 Mhz detecting longitudinal spin waves forming BECs. The phase delay data was taken at a fixed distance of .1 meters between the source coil and the detecting coil. These velocities easily exceed the speed of sound and these spin waves seem to be ubiquitous regardless of the metal. [67]**

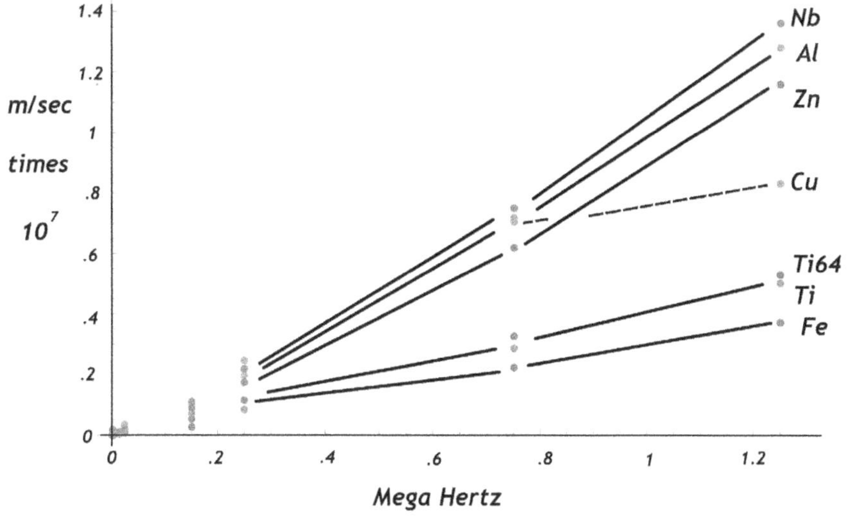

Understanding superconductivity was undermined by a fixation that lattice vibrations played a role in coupling electron pairs moving at much greater velocities than the speed of sound. The use of lattice vibration in the BCS theory was just an offshoot of using the simple harmonic oscillator from radiation theory and making a mistake about the source of the isotope mass effects. BCS theory failed for superconductivity, since coupling electrons at the fermi surface requires an excitation that both has a wide range of velocities and a mechanisms to bind electrons, where lattice vibrations can do neither. It turns out spherical longitudinal spin waves have the desired characteristics to accomplish this feat [67]. A single spherically propagating longitudinal spin wave can capture a pair of electrons with opposite propagation vectors if the spin wave velocity matches the electron velocities and then magnetic coupling can occur. The thermal longitudinal spin wave have a wide range of velocities as shown in Figure 3.9.

3.2   Debris

Early mistakes introduced into quantum mechanics in the 1920s were not corrected in the 1930s and thus crippled the field. The Standard Model of Particle Physics follows directly from Dirac's high energy approximation $E = pc$ where the self-energy is initially drop and then tacked on later. This slight of hand destroyed any understanding and continued to hide the incomplete state of special relativity. The argument mathematically inclined theorist like to use is that the Standard Model works so well it must be correct. It is the same argument used with respect to the Lamb shift correction from quantum electrodynamics and for the hydrogen atom solutions by the Schrödinger and Dirac equations.

3.2.1   Gauge Bosons & the Higgs

The gauge boson, single virtual photon, and renormalization were used to disguise the point charge that is the erroneous assumption buried in the Standard Model [35]. The most ridiculous application of a gauge boson is for the weak force where it was assumed that the 80 GeV $W^{\pm}Z^0$ particles are the intermediate force carriers and not the structures of the particles themselves that control the weak transitions. Modern high energy theory challenged Fermi's thought that the weak force is simply a form of a phase transition [69].

The basket of colorful gluons and quarks that were used to build the baryons and mesons are an unnecessary complication. The one and two dimensional wave functions that are solutions to self-reference frame wave equation are not points of matter just as the electron is not a point. When a high energy electron scatters off a proton the first excitation threshold produces a spin zero neutral massive boson, pion. The spin zero pion like the Higgs, neutral kaon, and longitudinal spin wave are nature's collectors of excess energy that needs to be safely dissipated. The charge based excitation of a proton in an accelerator experiment allows the proton to shed its excess energy at the threshold to form a neutral pion. With no dynamics in the self-reference the proton is barred from having a stable excited state. The Higgs field is theorized to give mass to all other particles except for themselves. That oxymoron is cured by the fact the particles themselves generate their own inertia with their own structure in their own self-reference frame.

Yukawa's Thesis of Nuclear Binding

Probably the most destructive theory that appeared in the 1930s was Yukawa's idea about nuclear binding. His model holds sway today in a different form called quantum chromodynamics (QCD) where the exchange boson is no longer the pion, but is now the gluon binding the individual pieces of the baryon. The complex process of virtual particle exchange can be supplanted by a simple contact interaction with components that have a spatial scale and are not points of matter. The binding energy for the contact interaction comes from the particle supplying the energy itself as found for nuclear matter by reducing their mass when expanding their scale.

Particle stability relies on a constant renewal process of creating a particle/anti-particle of the same type to generate its inertia. It is a process that does not require any excess information as the template already exists for the process in the original particle. The boson exchange process requires the creation of a different particle that alters the information content unlike the pair production process. The neutron has the added flexibility of being able to easily change its scale, mass, without altering its neutral charge to aid in nuclear binding [13].

3.2.2 Massive Neutrinos

The strange case of the missing solar neutrinos was based on neutrinos changing their flavor, type, because of a guess that if they were massive they could change their spots. The assumption was made that the electron neutrino's interaction cross section across the energy scale was equal to their creation probability. The neutrino as a massless field has half of the spatial density of the photon and hence its interaction cross section is reduced by one half and there are no missing solar neutrinos [58]. The neutrinos also are refracted by the earth's mass so that neutrinos passing through the earth are found in greater numbers with vertically oriented cylindrical detectors at midnight than when the sun is directly overhead. The refractive index can be deduced from this data and is compatible with the weak nuclear interaction [45]. Refraction is not a property of massive particles. The electron neutrino is the massless partner of the electron and results from a simple transformation applied to the electron.

3.2.3 Dark Matter

Zwicky's findings about the missing gravitational attractors from the bright matter in galaxies affecting rotation rates has been a long standing mystery. A diffuse gravitational mass that does not interact with radiation was a mystery until the discovery of the relativistic longitudinal spin waves. These longitudinal spin waves are massive spin 0 long lived boson excitations detected in Bose-Einstein condensates (BEC). These spin waves were first reported in iron with a scale of .14 meters in 2009 [23] [24] and more recently found with a scale of 1 meter or greater in a wide variety of metals and alloys [67]. These spin waves can exist in low density space operating with both electron and nuclear spins on a much greater scale and as massive bosons can collect in large numbers. A collection may have been imaged by x-rays scattering from dust attracted by their mass into a very large atomic like *2P* state forming the energy bubbles of the milky way [46].

3.2.4   Dark Energy

The dark motif was extended from matter to energy by neglecting the gravitational red shift of the photon. The reason for this was a poor understanding of the photon's structure as a propagating spherical shell in its own frame of reference that would be affected by all the mass this wave front would encompass from the time of its creation. The sum of this mass that this shell encompasses will reduce the frequency of the photon by a gravitation red shift [70]. In the intermediate distance region it looks much like the linear Hubble law, but at a greater distance the frequency is rapidly reduced mimicking the recent dark energy data, see Figure 2.2. Instead of dark energy it is simply a large volume gravitational effect that the photon is working against that is reducing its energy.

Questions

1) The neutron in a bound state can be stabilized when some of its mass is lost and it can no longer transform into a lower energy proton. Can a spin zero boson be stabilized in a Bose-Einstein condensate? Note mass plays a large role in the properties of a Bose-Einstein condensate.

2) The energy bubble straddling the Milky Way's central black hole appears as a large, $\sim 50,000$ light years, 2P atomic state. Why is there no 1S or 2S states? What would be the mass of an entity bound within such a large structure? Would these entities be optically visible?

3) The fracturing of rock can produces nuclear decay products, see *Piezonuclear neutrons from fracturing of inert solids* by F. Cardone, A. Carpinteri, and G. Lacidogna. This implies energy is being concentrated so that MeV collision are taking place. What two ways can energy be concentrated in the crushing of stone?

4) Can either of these energy concentration process be active in the atmosphere of the sun or in a volcano? What are implications for the chemistry of those two places?

5) If space is a conserved quantity, what are its astrophysical implications?

6) Why doesn't an electron, proton, or a neutron have an excited state in their self-reference frame?

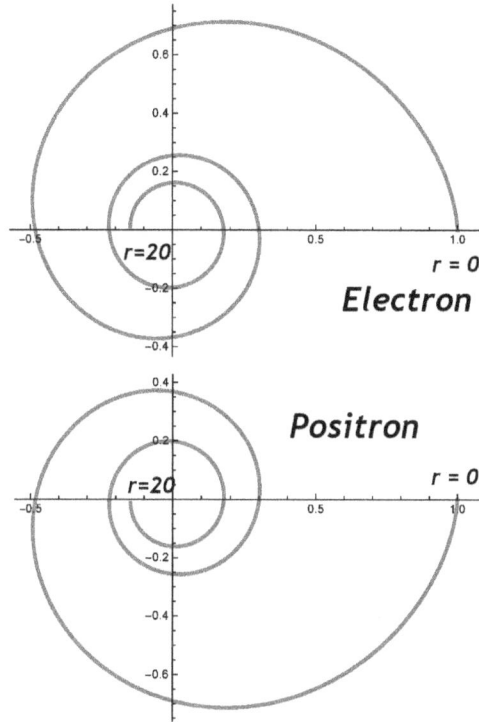

Figure 3.10: **Self-Reference frame plot of u(r) at relative rest where the electron ($\gamma = 1$) and positron ($\gamma = -1$) amplitudes plotted on the complex plane as a function of distance from their center of symmetry in their self-reference frames. The distance units are in terms of the Compton scale for the electron, $1/\kappa \ = 1 = \hbar/m_e c = 3.8 \times 10^{-13} meters$. γ is from special relativity and can take on values both less than one and zero and the dimension $n = 3$ see section 4.68 for the derivation. $_1F_1$ is a confluent hypergeometric function. [10]**

$$\mathbf{u}(r, \kappa, \gamma, n) \sim e^{-\kappa r} \, _1F_1\left[\frac{n-1}{1+i\gamma}, n-1, (1+i\gamma)\kappa r\right]; \quad n > 1$$

Chapter 4

Tool Box

The principal tool for investigating physics is its history from original sources. A broad exposure to the subject builds confidence in the consistency of arguments across different areas. The basic physical laws are expressed in differential equations that energy conservation yields. Practice in applying these tools to solving real problems is required. Here the hydrogen atom is used as an example when solved with the full relativistic equation, see Section 4.1.1.

Two press releases by Polykarp Kusch on the role of research institutes in universities and on education in general are used to start this chapter. These views are presented because they express ideas that were suppressed once the university became a victim of a faculty selected for their political bias rather than their accomplishments. This hiring trend after the 1960s led to the inability of the physics establishment to correct its major errors.

4.0.1   Kusch on Research Institutes

News Office Columbia University 29 March 1962
Summary of remarks by Polykarp Kusch at the Panel Discussion in the Shoreham Hotel, Washington, D.C., March 28, 1962 [48]

When this symposium was planned a few weeks ago, we had all hoped, I think, to find vigorous and perhaps instructive disagreement with each other. After having heard the earlier speakers I might as well have stayed in New York, a little better prepared than I will be to banish ignorance of the theory of electricity at 8:30 tomorrow morning.

I would like to be quite specific about a single aspect of our topic tonight. Barring only the accusation that a member of the academic community is not a scholar, no more

DAMNING STATEMENT CAN BE MADE ABOUT A COLLEAGUE THAN HE
IS ANTI-RESEARCH. WITH THIS KNOWLEDGE IN MIND, I WILL STILL
SUGGEST THAT THE UNIVERSITY OUGHT NOT UNDERTAKE EVERY POS-
SIBLE KIND OF RESEARCH, THAT IT OUGHT NOT TO UNDERTAKE MUCH
OF THE RESEARCH THAT CAN AND IS, BY AND LARGE, DONE QUITE AS
EFFECTIVELY IN OTHER WAYS, THAT IT OUGHT NOT TO PUT SUCH A
PREMIUM ON RESEARCH THAT WHAT MR. BARZUN HAS CALLED "THE
TRADITIONAL PURPOSE" IS WEAKENED. I RECOGNIZE THE NEED OF
WELL-DEFINED ADMINISTRATIVE ENTITIES, AS DEPARTMENTS, IN THE
UNIVERSITY AS ELSEWHERE, TO MANAGE THE AFFAIRS OF A GROUP
OF PERSONS WITH SIMILAR PURPOSES AND GOALS. TO THE DEGREE
THAT THE INCREASINGLY COMMON TERM "RESEARCH INSTITUTE" DE-
SCRIBES AN ENTITY WITH PURPOSES WHICH NOT ONLY DIFFER FROM
THOSE DESCRIBED BY MR. BARZUN, BUT WHICH WEAKEN THEM AS
WELL, I AM OPPOSED TO RESEARCH INSTITUTES.

ALL THIS MAY SOUND A BIT ODD IN A CITY FROM WHICH MANY, IF
NOT ALL, OF THE BLESSINGS OF RESEARCH FLOW AND IN THE MOUTH
OF THE PROFESSOR WHO BELONGS TO A CLAN THAT HAS BEEN ENTER-
PRISING AND SUCCESSFUL IN GAINING SUPPORT FOR RESEARCH.

I THINK THAT UNIVERSITY RESEARCH SHOULD BE DONE IN AN AT-
TEMPT TO ANSWER QUESTIONS OF GENERALITY AND IMPORTANCE. I
GRANT THE DIFFICULTY OF DEFINING THE WORDS GENERALITY AND
IMPORTANCE. STILL, THE MARK OF A GOOD QUESTION IS ONE THAT
MAY LEAD TO AN ANSWER APPLICABLE TO A LARGE RANGE OF HUMAN
EXPERIENCE, THAT WILL ANSWER QUESTIONS NOT YET ASKED AND
BARELY POSSIBLE OF BEING ASKED. THE MEMEBERS OF THE UNIVER-
SITY ATTEMPT TO COPE WITH THE EXPLOSION OF KNOWLEDGE. IN MY
OWN FIELD, PHYSICS, THERE IS INDEED SUCH AN EXPLOSION. STILL,
THE FIRST THREE DECADES OF THIS CENTURY WERE INCOMPARABLY
GREATER YEARS FOR PHYSICS THAN THE SECOND THREE DECADES. A
WHOLE NEW WELTANSCHAUUNG WAS DEVELOPED IN THE EARLIER
DECADES; THE LATER ONES ARE MARKED BY REFINEMENTS, NEW IN-
SIGHTS AND VAST BODIES OF NEW DATA. I THINK THE HISTORIAN OF
MY SUBJECT, SOME CENTURIES FROM NOW WILL ASCRIBE GREATNESS
ONLY TO THE EARLIER DECADES.

THE POINT I WISH TO MAKE IS THAT THE ACCUMULATION OF
MASSES OF REPORTS DOES NOT NECESSARILY MAKE FOR GREATNESS.

I REPEAT THEN, THAT UNIVERSITY RESEARCH SHOULD BE CON-
SEQUENT TO THE ASKING OF PROFOUND QUESTIONS. THE QUESTION
SHOULD BE " WHAT, OF THE WORLD AND OF THE MEN THAT INHABIT
IT AND OF THEIR INTERRELATIONSHIPS DO I NOT UNDERSTAND?" IT
SHOULD NOT BE "WHAT CAN I DO TO PARTICIPATE IN THIS MAR-
VELOUS ACTIVITY CALLED RESEARCH?"

THE ACQUISITION AND ANALYSIS OF DATA, SCIENTIFIC, ECONOMIC
OR SOCIAL, TO MEET IMMEDIATE PRACTICAL NEEDS IS A USEFUL FUNC-
TION OF ALL KINDS OF ORGANIZATIONS. CERTAINLY INDUSTRIAL
ENTERPRISE WITH ITS LARGE LABORATORIES PRODUCES ENORMOUS
VOLUMES OF TECHNICAL DATA. I DO NOT THINK THAT THE UNI-
VERSITY SHOULD TRY TO COMPETE WITH INDUSTRY IN THIS PROCESS.
VERY IMPORTANTLY, THE UNIVERSITY SHOULD NOT EXPEND ITS RE-
SOURCES, PHYSICAL, INTELLECTUAL OR EMOTIONAL IN COMPETING
WITH INDUSTRY TO PRODUCE A HUGE VOLUME OF TECHNICAL DATA.
THE UNIVERSITY SHOULD USE ITS RESOURCES IN AN EXPLORATION
OF WHAT IS CONCEPTUALLY NEW, IN AN ATTEMPT TO CREATE IN-
TELLECTUAL CONSTRUCTS OF LARGE DIMENSIONS. THE DETAILED,
THE PARTICULARIZED, THE APPLICATION OF IDEAS TO THE PRODUC-
TION DEVICES AND THE DEVELOPMENT OF TECHNIQUES ARE ALL IN
THE REALM OF WHAT IS APPROPRIATE AND EFFECTIVELY DONE ELSE-
WHERE THAN AT THE UNIVERSITY.

THE CONTEMPORARY BELIEF IN THE VALUE OF ANYTHING AT ALL
THAT CAN BE DESCRIBED AS RESEARCH HAS PUT A HIGH PREMIUM ON
WHAT IS OFTEN TRIVIAL, UNIMPORTANT AND DULL. THE BELIEF CAN
IMPAIR A BELIEF IN THE IMPORTANCE OF TEACHING, HOWEVER YOU
WISH TO DEFINE TEACHING. IT CAN IMPAIR THE TASTE OF THE ACA-
DEMIC COMMUNITY IN INTELLECTUAL MATTERS. IT CAN DISSIPATE
THE STRENGTH OF THE UNIVERSITY IN THE PURSUIT OF PURPOSELESS
FASHION.

I QUOTE SEVERAL TITLES FROM A BULLETIN OF THE <u>AMERICAN
PHYSICAL SOCIETY</u>:

"MICROWAVE ROTATIONAL SPECTRUM OF BENZALDEHYDE"

"OPTICAL AND ELECTRON SPIN RESONANCE SPECTRA OF AMMO-
NIUM
     VANADYL TARTRATE"

"ENERGY RESOLUTION IN THE STACKED-FOIL DETECTION OF HEAVY IONS"

I AM NOT A CRITIC OF THE CONTENT OF THE PAPERS. STILL, IF THE PRODUCTION OF THESE PAPERS KEPT SOMEONE FROM PERCEPTIVELY EXPLORING THE KNOWLEDGE OF THE SUB-MICROSCOPIC WORLD THAT PHYSICS HAS BROUGHT WITH STUDENTS, WHETHER FRESHMEN OR GRADUATE STUDENTS, THE PRODUCTION OF THE PAPERS IS A POSITIVE LIABILITY AND NOT A FULFILLMENT OF THE PURPOSES DESCRIBED BY MR. BARZUN.

I HOPE TO BE ALLOWED A FEW MORE WORDS ABOUT RESEARCH INSTITUTES. I DO NOT OBJECT TO THE VERBALISM. TO MY MIND RESEARCH INSTITUTES TEND TO DESCRIBE GROUPS THAT ARE PARTLY DETACHED FROM THE OVERALL LIFE OF THE UNIVERSITY. THEY TEND TO INQUIRE INTO MATTERS THAT ARE VERY SPECIFIC, RATHER THAN GENERAL. THEY TEND TO WIDEN THE GAPS IN HUMAN KNOWLEDGE, TO ENCOURAGE ITS FRAGMENTATION RATHER THAN TO CLOSE GAPS AND CONSOLIDATE KNOWLEDGE. THEY TEND TO BE COMMITTED TO LINES OF INQUIRY AND NOT TO BE UNCOMMITTED. THEY ATTRACT THE ALLEGIANCE OF THE UNIVERSITY MEN TO A PARTICULAR SUB-ENTITY WITHIN THE UNIVERSITY RATHER THAN TO LARGE ENTITIES WITHIN THE UNIVERSITY AND TO THE UNIVERSITY ITSELF. THEY ENCOURAGE THE BELIEF THAT THE MOST IMPORTANT THING IN THE UNIVERSITY IS THE RESEARCH OF THE INSTITUTE RATHER THAN THE BELIEF THAT THE RESEARCH IS AN IMPORTANT INGREDIENT OF THE UNIVERSITY. IN THE SENSE IN WHICH THEY DO ALL THESE THINGS, I THINK THE UNIVERSITY OUGHT NOT TO CULTIVATE RESEARCH INSTITUTES.

4.0.2   Kusch on Teaching

POLYKARP KUSCH'S THOUGHTS ON TEACHING PRESENTED IN DETROIT, MICHIGAN FEBRUARY 8, 1960 [71].

IT HAS BECOME THE FASHION CONDUCT SYMPOSIA THAT DEAL WITH ONE ASPECT OR ANOTHER OF THE PROBLEMS THAT CONFRONT EDUCATION IN AMERICA TODAY. THE NEED OF A CRITICAL, INFORMED AND HONEST APPRAISAL OF OUR SYSTEM OF EDUCATION IS NOT A

CONTROVERSIAL ISSUE. EVEN THE BEST THAT WE CAN DO TO EDU-
CATE THE YOUNG MAY WELL NOT BE GOOD ENOUGH TO ALLOW THEM
TO CARRY ON THE TRADITION OF OUR LIBERAL WESTERN CIVILIZA-
TION IN THE FACE OF PROBLEMS OF UNPRECEDENTED COMPLEXITY.
STILL, WE WOULD NOT BE KEEPING FAITH WITH THAT TRADITION
WERE WE NOT TO TRY TO EXCEED WHAT WE MAY WELL HAVE BE-
LIEVED TO BE OUR BEST.

IT APPEARS TO ME THAT THE GRAVEST PROBLEMS OF OUR PRESENT
EDUCATIONAL SYSTEM ARISE NOT SO MUCH WITHIN THE ACADEMY IT-
SELF AS FROM THE SOCIAL AND CULTURAL ENVIRONMENT IN WHICH
THE ACADEMY EXISTS. THE PRESENT ENVIRONMENT HAS, I THINK,
A STRONG INHIBITING EFFECT ON THE DIFFUSION OF KNOWLEDGE,
UNDERSTANDING AND PERCEPTION AT ALL LEVELS OF SOCIETY. AN
AUTHORITATIVE ESTIMATE OF THE ADVERSE EFFECT OF THE ENVIRON-
MENT ON THE EDUCATIONAL PROCESS WOULD OCCUPY ONE FOR A
VERY LONG TIME; I WILL, HERE, COMMENT ON ONLY ONE ASPECT OF
THE SOCIAL FRAMEWORK IN WHICH EDUCATION OCCURS.

THERE IS A CULT IN AMERICA THAT ASSERTS THAT KNOWLEDGE
AND UNDERSTANDING MAY BE MADE EASY TO ACQUIRE. THE SYM-
BOLS OF THE CULT ARE MANY — POPULAR MAGAZINES THAT PURPORT
TO PRESENT IN A BRIEF CONDENSATION THE ESSENCE OF A LONG AND
THOUGHTFUL STUDY OF, SAY, THE HUMAN CONDITION — ARTICLES IN
THE PICTURE MAGAZINES THAT PURPORT TO GIVE THE LOW-DOWN
ON, SAY, NUCLEAR PHYSICS — A PROLIFERATION OF BOOKS WITH TI-
TLES LIKE "MATH MADE EASY." THESE PRESENTATIONS MAY WELL
SERVE A USEFUL PURPOSE BUT IT IS DANGEROUS TO SUPPOSE THAT
THEY GIVE AN EDUCATION OR EVEN A SIGNIFICANT PART OF ONE.
THE BELIEF THAT LEARNING MAY BE MADE EASY HAS INFECTED THE
ACADEMY — THE SECONDARY SCHOOLS, THE COLLEGES, THE GRADU-
ATE SCHOOLS — AND THIS BELIEF HAS LED, I THINK, TO DIMINISHING
DEMANDS ON OUR STUDENTS THAT IS, THE BELIEF THAT LEARNING IS
EASY HAS LED US TO OFFER A KIND OF LEARNING THAT IS, IN FACT
EASY.

EXCEPT, PERHAPS, FOR A GIFTED FEW, MATHEMATICS IS NOT EASY,
NOR IS NUCLEAR PHYSICS, NOR, FOR THAT MATTER, IS ANY OTHER
SUBJECT OF GREAT INTELLECTUAL CONTENT, AS THE STUDY OF THE
HUMAN CONDITION, IN ITS PAST AND PROSPECTS. PERHAPS INTEL-

LECTUAL ENDEAVOR IS EASY EVEN FOR THE GIFTED FEW; THESE, TO-
GETHER WITH A SUPERIOR INTELLECTUAL ENDOWMENT, MAY WELL
HAVE THE GIFT OF A SELF-IMPOSED DISCIPLINE. I DOUBT THAT ANY-
ONE HAS EVER ACQUIRED SIGNIFICANT UNDERSTANDING OF ANYTHING
BY TURNING ON THE RADIO, SETTLING IN AN EASY CHAIR, AND READ-
ING A CONDENSATION OF WISDOM IN THE SPIRIT IN WHICH A DETEC-
TIVE STORY MIGHT BE READ. TRUE UNDERSTANDING IS ACHIEVED
ONLY BY HARD AND EXHAUSTING MENTAL EXERCISE, BY MEN AND
WOMEN WHO ARE IN TRAINING FOR SUCH EXERCISE. THE CULT OF
EASINESS IMPAIRS A RESPECT FOR THE MAGNIFICENT INTELLECTUAL
STRUCTURE OF, SAY, PHYSICS AND DOES NOT ALLOW FOR THE CULTI-
VATION OF THE INVALUABLE KNOWLEDGE THAT WHATEVER UNDER-
STANDING WE HAVE OF NATURE AND OF MAN HAS BEEN OBTAINED
BY ARDUOUS TOIL. TRUE KNOWLEDGE GIVES AND ENCOMPASSING
UNDERSTANDING OF THE WHOLE, AN APPRECIATION OF CAUSES AND
CONNECTIONS, OF CONCEPTUAL MODELS OF NATURE AND THEIR RE-
LATIONSHIP TO OBSERVABLE REALITY. THE ACQUISITION OF THIS KIND
OF KNOWLEDGE IS HARD WORK, PERHAPS AS HARD AS ANY WORK
KNOW TO MAN AND IT MAY OFTEN BE FRUSTRATING.

IF WE ARE TO PRODUCE AND MAINTAIN AN INTELLECTUAL ELITE,
THE DIFFICULTY OF LEARNING IS ONE IDEA THAT MUST BE NURTURED.
NO MAN, NO WOMEN, SHOULD HAVE THE ACADEMY WITHOUT THE
ABIDING CERTAINTY THAT THE KNOWLEDGE THAT HAS BEEN WON
HAS BEEN PAINFULLY WON, AND PERHAPS TRIUMPHANTLY WON, PRE-
CISELY BECAUSE THE WINNING WAS PAINFUL. OUT OF SUCH PER-
SONAL KNOWLEDGE WILL COME A RESPECT FOR ALL LEARNING AND
ALL KNOWLEDGE. THE EFFECTIVENESS OF OUR EDUCATION WILL BE
ENORMOUSLY ENHANCED IF THE STUDENT ACQUIRES NOTHING MUCH
MORE THAN THE INSTINCT FOR RESPECT OF LEARNING, AN ADMIRA-
TION FOR IT QUITE AS GREAT AS THAT PRESENTLY RESERVED FOR
OTHER DEMANDING HUMAN ACTIVITIES, LIKE MAKING MONEY AND
PLAYING BASKETBALL.

WHAT I HAVE SAID IS, PERHAPS, IN THE NATURE OF PIOUS PLAT-
ITUDES, A SORT OF DENUNCIATION OF GENERALIZED SIN. I SUG-
GEST THAT WE SHOULD GO MUCH FURTHER THAN MERELY DENOUNC-
ING SIN; WE MUST MAKE HARD AND RIGOROUS DEMANDS OF OUR
STUDENTS AT ALL LEVELS. THERE HAS BEEN, I THINK, A CONTIN-

UING RELAXATION OF THE ENTRANCE DEMANDS MADE BY ADMIS-
SION OFFICERS IN OUR COLLEGES AND UNIVERSITIES. THE PROBLEMS
THAT FACE US BECOME MORE COMPLEX, THE LEARNING TO BE MAS-
TERED MORE EXTENSIVE. THE ENTERING STUDENT SHOULD HAVE A
GREATER, NOT A LESSER COMMAND OF FOREIGN LANGUAGES, MATH-
EMATICS, AND SCIENCE. MOST ESPECIALLY, HE SHOULD BE ABLE TO
USE THE ENGLISH LANGUAGE NOT ONLY WITH ACCURACY BUT ALSO
WITH GRACE. I THINK THAT IF WE IMPOSED THESE DEMANDS, THESE
DIFFICULT DEMANDS, THEY WOULD ULTIMATELY BE MET. I FEEL THAT
NO STUDENT SHOULD RECEIVE A BACCALAUREATE DEGREE WITHOUT
A FINISHED ABILITY TO USE A FOREIGN LANGUAGE, WITHOUT A DE-
VELOPED SKILL IN THE THE USE OF THE QUANTITATIVE METHODS
THAT CHARACTERIZE MATHEMATICS AND CERTAIN OF THE SCIENCES.
I THINK THAT WERE WE REALLY TO INSIST ON EXCELLENCE, WERE WE
TO DEMAND HARD AND PERHAPS UNREMITTING WORK, WE WOULD BE
LESS CONCERNED BY THOSE STUDENTS THAT JUST CAN NOT DO MATH-
EMATICS, USE A FOREIGN LANGUAGE OR WRITE GOOD ENGLISH. WE
WOULD BE LESS CONCERNED SIMPLY BECAUSE THERE WOULD BE LESS
OF THESE PROBLEM CASES. I THINK THAT NO ONE WHO CARRIES THE
TITLE DOCTOR OF PHILOSOPHY SHOULD BE LESS THAN A PHILOSO-
PHER, A MASTER OF SOME KNOWLEDGE AND AN EAGER APPRENTICE
OF ALL KNOWLEDGE.

I URGE THAT EDUCATION BE MADE DIFFICULT AND DEMANDING,
THAT THE PAINSTAKING EFFORT REQUIRED TO LEARN BE RECOGNIZED
AS AN IMPORTANT INGREDIENT OF KNOWLEDGE. I CANNOT BELIEVE
THAT AN EDUCATION DESIGNED TO ENCOMPASS AN EVER INCREAS-
ING FRACTION OF THE POPULATION BY MAKING LEARNING AN EVER
EASIER PROCESS CAN DO ANYTHING BUT SPELL DISASTER FOR US.

4.1   Basic Findings

Referenced are results from different experimental areas that
are consistent with the revised differential equations of quantum mechanics. The abstracts of these works are included along
with the URLs to the complete papers. More extensive and
detail derivations, particular for relativistic behavior will be

found in *"yes Virginia", quantum mechanics can be understood 3rd edition* and *Principles of Matter amending quantum mechanics*. We start with an experiment started by Kusch to prove that quantum electrodynamics was invalid.

4.1.1 Hydrogen Atom's Ground State

EXTRACTED FROM [13]

IN 1947 POLYKARP KUSCH AND HENRY FOLEY REPORTED THEIR EXPERIMENT OF THE GYROMAGNETIC RATIO OF THE ELECTRON [41]. THIS FINDING WAS EVENTUALLY TITLED BY OTHERS AS THE ANOMALOUS MAGNETIC MOMENT, BEING A CORRECTION TO THE MAGNETIC MOMENT DERIVED FROM THE DIRAC EQUATION. THERE WAS AN INTENSE EFFORT BOTH EXPERIMENTAL AND THEORETICAL TO ESTABLISH THE SPECTROSCOPIC STRUCTURE OF THE HYDROGEN ATOM THROUGH OUT THE 1930s AND 1940s [72]. THIS WORK ON THE MAGNETIC MOMENT IN PART GAVE RISE TO A THEORY CALLED QUANTUM ELECTRODYNAMICS THAT WAS USED TO COMPUTE THE CORRECTIONS TO THE DIRAC ELECTRON THEORY. UNTIL THE EARLY 1960s P. KUSCH HAD NO PUBLIC PROBLEM WITH QUANTUM ELECTRODYNAMICS BUT BY 1967 HE HAD COME TO THE CONCLUSION IT WAS NOT VALID PHYSICS. THE TROUBLING EXPERIMENTAL QUESTION WAS THE DISCREPANCY IN THE GROUND STATE ENERGY OF HYDROGEN $^2S_{\frac{1}{2}}$. THE EXPERIMENTAL DISCREPANCY TO BOTH THE SCHRÖDINGER AND DIRAC EQUATIONS COMPUTED VALUES WERE LARGE, $> .003$ EV. TO A SPECTROSCOPIST THIS IS A LARGE DISCREPANCY. TO CORRECT FOR THIS DISCREPANCY THE FULL RELATIVISTIC WAVE EQUATION SOLUTION IS REQUIRED.

$$i\hbar\frac{\partial\Phi}{\partial t} = -\frac{\hbar^2}{m_0(1+\gamma)}\nabla^2\Phi + \frac{2V}{1+\gamma}(1+\frac{V}{2m_0c^2})\Phi \quad (4.1)$$

THE TRIAL SOLUTION FOR THE TIME DEPENDENT PORTION OF THE WAVE FUNCTION:

$$\Phi(r,t) = \phi(r)e^{-\frac{iE_{rel}t}{\hbar}} \quad (4.2)$$

PRODUCING.

$$E_{rel}\ \phi(r) = -\frac{\hbar^2}{m_0(1+\gamma)}\nabla^2\phi + \frac{2V}{1+\gamma}(1+\frac{V}{2m_0c^2})\phi \qquad (4.3)$$

TAKING $\phi(r) = r^s e^{-\beta r}$ YIELDS IN 3D SPHERICAL COORDINATES:

$$E_{rel}\ \phi(r) =$$

$$-\frac{\hbar^2}{m_0(1+\gamma)}\{\beta^2 - \frac{2\beta(s+1)}{r} + \frac{s(s+1)}{r^2}\}\phi(r)\ + \qquad (4.4)$$

$$\frac{2V}{1+\gamma}(1+\frac{V}{2m_0c^2})\phi(r)$$

APPLYING THE COULOMB POTENTIAL TO THE TWO POTENTIAL TERMS THE RELATIVISTIC WAVE EQUATION BECOMES:

$$V + \frac{V^2}{2m_0c^2} = -\frac{Ze^2}{2(1+\gamma)\pi\epsilon_o r} + \frac{1}{(1+\gamma)m_0c^2}\frac{Z^2e^4}{(4\pi\epsilon_o)^2 r^2} \qquad (4.5)$$

SEPARATING THE TERMS IN POWERS OF R ALLOWS E_{rel}, s AND β TO BE COMPUTED AS ALL FACTORS OF r^m MUST EQUAL ZERO.

$$E_{rel} + \frac{\hbar^2\beta^2}{m_0(1+\gamma)} =$$

$$\frac{1}{r}\{-\frac{2\hbar^2\beta(s+1)}{m_0(1+\gamma)} - \frac{Ze^2}{2(1+\gamma)\pi\epsilon_o}\}\ + \qquad (4.6)$$

$$\frac{1}{r^2}\{-\frac{\hbar^2 s(s+1)}{m_0(1+\gamma)} + \frac{1}{(1+\gamma)m_0c^2}\frac{Z^2e^4}{(4\pi\epsilon_o)^2}\}$$

SOLVING FOR s AND THEN SUBSTITUTING FOR THE FINE STRUCTURE CONSTANT α:

$$s^2 + s = \frac{Z^2e^4}{(4\pi c\hbar\epsilon_o)^2} = \alpha^2 Z^2 \qquad (4.7)$$

$$s = \sqrt{1 + 4Z^2\alpha^2} - 1 \qquad (4.8)$$

For β the equation is simplified by using the Bohr radius a_0:

$$\beta = \frac{1}{s+1}\frac{m_0}{\hbar^2}\frac{Ze^2}{4\pi\epsilon_0} = \frac{Z\alpha}{\sqrt{1+4Z^2\alpha^2}}\frac{m_0 c}{\hbar} = \frac{Z}{a_0\sqrt{1+4Z^2\alpha^2}} \quad (4.9)$$

Then the expression for the ground state energy of the $^2S_{\frac{1}{2}}$ state when the factor representing the reduce mass effect is applied where m_N is the nucleon mass. [64].

$$E_{rel} = -\frac{m_0 c^2}{(1+\gamma)}\frac{Z^2\alpha^2}{1+4Z^2\alpha^2}\frac{m_N}{m_0+m_N} \quad (4.10)$$

The γ dependence can be factored out by using the expression for total energy $E_t = \gamma m_0 c^2$.

$$E_{rel} = E_t - m_0 c^2 = m_0 c^2(\gamma - 1) \quad (4.11)$$

Substituting into equation 4.10 to remove γ yields:

$$E_{rel}^2 + 2m_0 c^2 E_{rel} - 4(m_0 c^2)^2\frac{Z^2\alpha^2}{1+4Z^2\alpha^2}\frac{m_N}{m_0+m_N} = 0 \quad (4.12)$$

$$E_{rel} = -m_0 c^2\{-1 + \sqrt{1 + \frac{Z^2\alpha^2}{1+4Z^2\alpha^2}\frac{m_N}{m_0+m_N}}\} \quad (4.13)$$

The wave function then becomes:

$$\Phi(r,t) = Ar^{\sqrt{1+4Z^2\alpha^2}-1}\,e^{-\frac{Zr}{a_0\sqrt{1+4Z^2\alpha^2}}-i\frac{Et}{\hbar}} \quad (4.14)$$

The ground state energies for the three different wave equations are computed in Table 3.1. γ in the bound state now takes on values less than one as the self-energy is reduced to supplying the field that binds the state. The change in γ can be computed using equation 4.11. When considering the solutions for the higher Z hydrogen like ions both the Schrödinger and Dirac equation show non-physical behavior when compared to the full relativistic solutions [13].

THE OTHER FEATURE THAT IS REVEALED IS THAT FOR THE ELECTRO-
STATIC POTENTIAL THERE IS A LIMIT TO THE BINDING STRENGTH OF
LESS THAN 10% OF THE SELF-ENERGY.

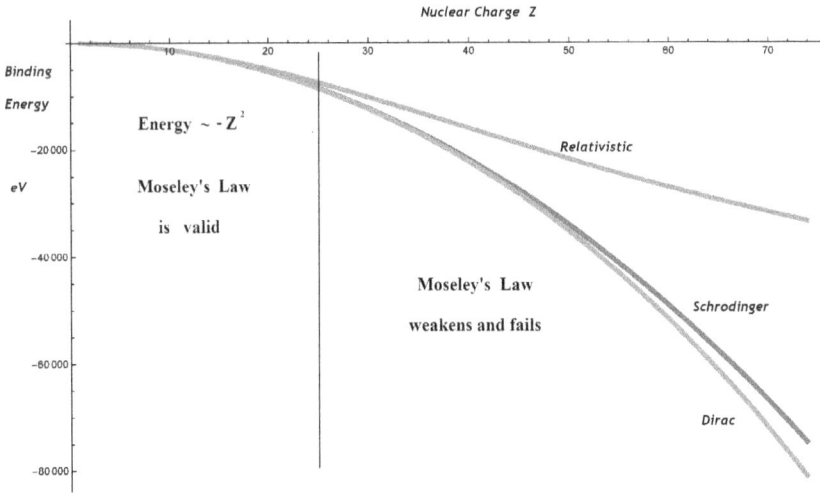

Figure 4.1:   **Binding energy of hydrogen like ions taking Z from
1 to 74 in eV for three models [13]. The break down in the Z^2
behavior for high values of Z does not fit the behavior of either the
Dirac or Schrödinger results for the K-shell ionization data.**

4.1.2   Spin

EXTRACTED FROM $[22]$

SPIN IS A DYNAMICS PROPERTY GENERATED IN THE LABORATORY
FRAME. THE FREE PARTICLE WAVE EQUATION FOR THE LABORATORY
FRAME THAT IS THE POTENTIAL FREE PRODUCES A SECOND ORDER
DIFFERENTIAL EQUATION IN SPACE AND TIME WITH A SOLUTION OF
THE FORM $\Phi(R,t) = W(R)G(t)$.

$$\{\nabla^2 - \frac{1}{c^2}\frac{\partial^2}{\partial t^2} + i\frac{2m}{\hbar}\frac{\partial}{\partial t}\}\Phi(R,t) = 0 \qquad (4.15)$$

SOLVING THE EQUATION STARTS BY SEPARATING THE TIME AND SPACE
DEPENDENT PORTIONS WITH THE SEPARATION CONSTANT η.

$$\frac{\nabla^2 W}{W} = \frac{1}{G(t)}\{\frac{1}{c^2}\frac{\partial^2 G(t)}{\partial t^2} - i\frac{2m}{\hbar}\frac{\partial G(t)}{\partial t}\} = \eta \qquad (4.16)$$

THERE IS NO QUALIFICATION ASSOCIATED WITH THESE EQUATIONS AS TO WHAT FAMILIES OF PARTICLE THEY CAN BE APPLIED, NO RESTRICTION AS SPIN OR ANY ASSIGNMENT TO A PARTICULAR PARTICLE SUCH AS REQUIRED BY THE KLEIN-GORDON AND DIRAC EQUATIONS.

$$\frac{\partial^2 G}{\partial t^2} + i\frac{2mc^2}{\hbar}\frac{\partial G}{\partial t} = c^2\eta G \qquad (4.17)$$

TAKING ONLY THE STABLE SOLUTION FROM TIME DEPENDENT EQUATION HAS A SIMPLE HARMONIC TIME DEPENDENCE $G(t) = e^{i\omega t}$. THE SPATIAL EQUATION IN 3D SPHERICAL CASE THEN GENERATES A PAIR OF SOLUTIONS:

$$\nabla^2 W - \eta W = 0 \qquad (4.18)$$

$$W(R) = A\frac{e^{-\kappa R}}{R} + B\frac{e^{\kappa R}}{R} \qquad (4.19)$$

THE TERM THAT GROWS WITH INCREASING R IS NOT A PHYSICAL SOLUTION AND THEREFORE $B = 0$. THE TOTAL SOLUTION IS:

$$W(R)G(t) = A\frac{e^{-\kappa R - i\omega t}}{R} \qquad (4.20)$$

NORMALIZING THE WAVE FUNCTION THE VALUE OF A CAN BE DETERMINED.

$$1 = \int_0^\infty W^*G^*WG\, R^2 dV = A^*A4\pi\int_0^\infty e^{-2\kappa R}dR = \frac{2\pi A^2}{\kappa} \qquad (4.21)$$

$$A = \sqrt{\frac{\kappa}{2\pi}} \qquad (4.22)$$

$$\Phi(R,t) = W(R)G(t) = \sqrt{\frac{\kappa}{2\pi}}\frac{e^{-\kappa R - i\omega t}}{R} \qquad (4.23)$$

THE PARTICLE DENSITY FUNCTION IN THE REST STATE, $\rho(R)$ IS:

$$\rho(R) = 4\pi \; W^*G^*WG \; R^2 = 2\kappa \; e^{-2\kappa R} \qquad (4.24)$$

THE PARTICLE'S MEAN SCALE $< R >$ CAN NOW BE COMPUTED.

$$< R > = \int_0^\infty W^*G^* \; R \; WG \; R^2 dV =$$

$$(4.25)$$

$$2\kappa \int_0^\infty Re^{-2\kappa R} \; dR = \frac{1}{2\kappa} \int_0^\infty xe^{-x} \; dx = \frac{1}{2\kappa} = \frac{\epsilon}{2}$$

THIS PARTICLE SCALE OF $\epsilon/2$ IS ONE HALF OF THE RANDOM SCALE FROM THE COMPTON RELATION. THIS SCALE CAN BE APPLIED TO THE DYNAMICS OF ANGULAR MOMENTUM.

THE ANALYSIS BEGAN [22] WITH FIELDS TRAVELING AT THE SPEED OF LIGHT. TO DEFINE INERTIAL MASS PARTICLES HAVE A SCALE DE-TERMINED BY RANDOMIZING THE FIELD THROUGH A PAIR-PRODUCTION AND ANNIHILATION PROCESS. HAVING A PARTICLE WITH A SCALE MAKES DEFINING ANGULAR MOMENTUM SIMPLE. ANGULAR MOMEN-TUM IS THEN A PRODUCT OF THREE TERMS MASS, ITS VELOCITY, SPEED OF LIGHT, AND THE PARTICLE'S SCALE. THE SPHERICAL PARTICLE'S SYMMETRY ALLOWS ALL ACTION TO BE COMPUTED FROM THE CENTER OF SYMMETRY. THE DIFFICULTY OF DEALING WITH THE RELATIVIS-TIC CONTRACTION OF A ROTATING MASSIVE SOLID DISAPPEARS. THE PRODUCT REPRESENTS THE ANGULAR MOMENTUM. THIS IS NOT AN ORBITAL ANALYSIS, BUT RANDOM QUANTUM ACTION MAINTAINING A FINITE ANGULAR MOMENTUM OF A STABLE PARTICLE. BECAUSE OF ENERGY CONSERVATION THE RANDOM PROCESS THAT GENERATES IN-ERTIAL MASS ALSO FIXES A STABLE PARTICLE'S INTRINSIC ANGULAR MOMENTUM. THIS PRODUCT PRODUCES A SINGLE RESULT AN AN-GULAR MOMENTUM MAGNITUDE OF $\hbar/2$, WHICH TURNS OUT TO BE THE ANGULAR MOMENTUM OF STABLE FERMIONS THAT ARE THE ONLY STABLE MASSIVE PARTICLES.

$$L = mc < R > = mc\frac{\epsilon}{2} = mc\frac{\hbar}{2mc} = \frac{\hbar}{2} \qquad (4.26)$$

HAVING ONLY ONE ALLOWED VALUE OF THE ANGULAR MOMEN-
TUM INDICATES ANY OTHER VALUE FOR A MASSIVE PARTICLE MAKES
IT UNSTABLE. IN ORDER TO CAPTURE A PRIMITIVE RANDOMIZED FIELD
MOVING AT THE SPEED OF LIGHT ANGULAR MOMENTUM IS A NECES-
SARY PART OF THE PROCESS THAT STABILIZES THE FIELD AND ESTAB-
LISHES A LOCALITY FOR THE PARTICLE.

ANGULAR MOMENTUM DERIVED IN THIS WAY IS BOTH A SCALE
AND MASS INDEPENDENT PROPERTY. THE MAGNITUDE OF THE ANGU-
LAR MOMENTUM WILL BE THE SAME FOR AN ELECTRON AS WELL AS
A PROTON WITH THE DIFFERENCE IN THE TWO ALLOWED STATES OF
$+\hbar/2$ AND $-\hbar/2$ DEPENDING ON THE SIGN OF ROTATION. THIS DIF-
FERENCE ALLOWS RADIATIVE COUPLING TO PHOTONS WITH SPIN OF
$\hbar$. THERE IS NOTHING LINKING THE DIMENSIONAL DEPENDENCE TO
ANGULAR MOMENTUM OF THE PARTICLE INVOLVED EXCEPT THE SO-
LUTION IS DONE FOR THE THREE DIMENSIONAL REPRESENTATION OF
A MASSIVE PARTICLE. THE KEY PROPERTY OF ANGULAR MOMENTUM
IS ATOMIC, THAT OF THE WHOLE, NOT DENSITY DEPENDENT MAKING
ANGULAR MOMENTUM COMPATIBLE WITH AN ENSEMBLE OF NUCLEAR
COMPONENTS FORMED FROM A COLLECTION OF LOWER DIMENSIONAL
COMPONENTS.

Spin of a Massless Field

THERE ARE TWO MASSLESS FIELDS IN THREE DIMENSIONS THAT CAN
BE INVESTIGATED FOR THEIR STABLE VALUES OF ANGULAR MOMEN-
TUM, THE PHOTON AND ELECTRON NEUTRINO. THESE TWO FIELDS
ARE REPRESENTED BY EXPANDING SHELLS USING THE SOLUTIONS OF
THE RELATIVISTIC WAVE EQUATION THAT ARE FOUND IN SECTION 4.4.3.
STARTING WITH THE SAME ELEMENTARY EXPRESSION DEFINING THE
ANGULAR MOMENTUM USED FOR MASSIVE PARTICLES THAT MAY SEEM
OUT OF PLACE FOR A MASSLESS FIELD. THAT IS NOT THE CASE BE-
CAUSE OF HOW MASS IS TRANSFORMED FOR PRODUCING A MASSLESS
FIELD.

$$L = mc < R > \qquad (4.27)$$

THE EXPANDING FIELD DOES NOT HAVE A FIXED LOCATION AND THE
CONCEPT OF REST HAS NO MEANING FOR A MASSLESS FIELD RATHER IT

HAS A SCALE $<R>$. THE SHELL THICKNESS DEFINES THE DIMENSION OF THE WAVE FRONT AND THAT SCALE IS INVARIANT AS THE SHELL EXPANDS. HERE THE FIELD IS ONLY DEFINED OVER THE WIDTH OF THE WAVE FRONT $-i\epsilon$ BY TAKING A HINT FROM DIRAC'S WORK ON LIGHT FRONT [21]. THIS WAS THE CONDITION THAT INITIALLY GENERATED THE QUANTIZATION OF THE FIELD. THE EQUIVALENT MASS IS DEFINED FOR A MASSLESS FIELD IS A COMPLEX QUANTITY. THE FIELD REPRESENTATION IN THE RADIAL VARIABLE IN SPHERICAL COORDINATES IS GIVEN IN $W(R)$.

$$W(R) = \frac{1}{4\pi} \frac{e^{i\kappa R}}{R} \tag{4.28}$$

THE MEAN VALUE OF R CAN BE COMPUTED AS $<R>$ ON THE WAVE FRONT. WHERE THE TIME DEPENDENT TERMS YIELD A FACTOR OF 1.

$$<R> = \int W^* G^* R\, WGR^2 dR = \int_R^{R-i\epsilon} RdR = \tag{4.29}$$

$$\frac{(R-i\epsilon)^2}{2} - \frac{R^2}{2} = -i\epsilon - \frac{\epsilon^2}{2} = -i\epsilon$$

KEEPING ONLY FIRST ORDER TERMS IN, ϵ, WHICH ARE SMALL, THE SECOND ORDER TERMS IN, ϵ, ARE DROPPED. THE EFFECTIVE MASS LIKE $<R>$ IS ALSO COMPLEX.

$$m = -\frac{\hbar}{i\epsilon c} \tag{4.30}$$

THE PRODUCT SETTING THE ANGULAR MOMENTUM L IS:

$$L = -\frac{\hbar}{i\epsilon c}\, c\,(-i\epsilon) = \hbar \tag{4.31}$$

THE PHOTON HAS A DENSITY THAT IS UNITY, HOWEVER, THE NEUTRINO HAS A DENSITY OF ONE HALF WHEN COMPARED TO A PHOTON IN THE SELF-REFERENCE [22] [58]. BECAUSE OF THAT THE NEUTRINO'S ANGULAR MOMENTUM IS ONE HALF OF THAT OF A PHOTON.

$$L_{neutrino} = \frac{L_{photon}}{2} = \frac{\hbar}{2} \tag{4.32}$$

PAULI PRESSED THE CASE FOR USING THE KLEIN-GORDON EQUATION FOR MASSIVE SPIN ZERO PARTICLES AND DIRAC HAVING AN EQUATION SPECIFICALLY FOR THE ELECTRON BECAUSE IT PRODUCED A MAGNETIC MOMENT NEAR THE PHYSICAL VALUE WERE BOTH MISTAKES [34]. ANGULAR MOMENTUM MAKES UP ONLY A SMALL FRACTION OF A PARTICLE'S TOTAL SELF ENERGY [22]. SPIN'S ROLE IS MUCH GREATER RESTRICTING ONLY TWO ALLOWED VALUES FOR STABLE ENTITIES. THEORISTS WHO HAD USED THE GEOMETRY OF ROTATION AS THE PRIMARY GEOMETRY TO DEFINE SPACES MADE A SERIOUS MISTAKE THAT MISLED MANY STUDENTS [39].

4.1.3   Statistical Independence

EXTRACTED FROM [58]

THERE IS A MYTH THAT QUANTUM ELECTRODYNAMICS, A METHOD OF CALCULATION, HAS MADE QUANTUM MECHANICS THE MOST ACCURATE THEORY EVER. QUANTUM ELECTRODYNAMICS' NON-UNIQUE SET OF CORRECTIONS ARE CONSIDERED EVEN BY R. FEYNMAN, ONE OF THE ORIGINATORS, NOT A DESCRIPTION OF PHYSICS, BUT A METHOD OF CALCULATION. THE NON-UNIQUENESS ALLOWS RESULT TO USE NON-PHYSICAL PROPERTIES (POTENTIALS WITH SINGULARITIES) TO GENERATE ANY DESIRED NUMBER. WHEREAS, THE TWO EMPIRICAL CONSERVATION OF ENERGY RELATIONS ALLOWS ONE TO DERIVE A DESCRIPTION OF THE SPACE WHERE PARTICLE AND FIELD PROPERTIES ARE GENERATED, THE SELF-REFERENCE FRAME, AND THEIR SUBSEQUENT BEHAVIOR IN THE LABORATORY FRAME. THIS REQUIRES A TWO PART DERIVATION TO GENERATE BOTH STRUCTURE AND DYNAMICS IN TWO SEPARATE SPACES. THESE SPACES ARE NOT IN HIERARCHY, BUT COMPLIMENTARY WHERE EACH SPACE'S EXISTENCE IS DEPENDENT ON THE OTHER.

Self-Reference Frame

STARTING WITH THE MASSLESS FIELD CONSERVATION OF ENERGY AND RANDOMIZING MOTION FOR THAT FIELD BEGINS THE DERIVATION TO PRODUCE STRUCTURE AND INERTIA. THIS RANDOMIZING PROCESS IS GENERATED INDEPENDENTLY WHEN THE DYNAMIC EQUATION IS DE-

RIVED IN THE NEXT SECTION. THE PARTICLE'S STRUCTURAL FORM IN THE SPACE REFERENCED ON THE PARTICLE ITSELF CAN BE GENERATED BY TAYLOR EXPANDING THE MOMENTUM AND ENERGY OPERATORS AROUND r AND τ OF THE FIELD WITH THE RANDOM SPATIAL PARAMETER, ϵ AND TIME PARAMETER c/ϵ RESULTING IN TWO DIFFERENTIAL EQUATION, ONE FOR THE SPATIAL STRUCTURE, $u(r)$, AND ONE FOR THE TIME DEPENDENCE, $g(\tau)$, WHICH ARE DERIVED IN DETAIL [16] [35]. THE DERIVATION BEGINS WITH THE EXPANSIONS:

$$E = pc \quad \rightarrow \quad E = p(r+\delta r)c \quad \rightarrow \quad E = p(r+\epsilon)c$$
$$E = \hbar\omega \quad \rightarrow \quad E(\tau+\delta\tau) = \hbar\omega \quad \rightarrow \quad E(\tau+\frac{\epsilon}{c}) = \hbar\omega \tag{4.33}$$

THE ENTIRE CONCEPT OF A POINT MASS AND POINT CHARGE WITH THEIR ASSOCIATED INFINITE ENERGIES VANISH IN THIS PICTURE ALONG WITH THE CUT OFFS NECESSARY IN QUANTUM ELECTRODYNAMICS. WHAT ALSO VANISHES IS THE SINGLE VIRTUAL PHOTON, WHICH CANNOT BE SUPPORTED BECAUSE IT WILL CHANGE THE INFORMATION CONTENT OF THE LABORATORY FRAME.

STARTING WITH THE DISPERSION RELATION FOR A MASSLESS FIELD IN LABORATORY FRAME [35]:

$$p\ c = E \tag{4.34}$$

THE SPATIAL DEPENDENT EQUATION WILL BE DERIVED FIRST WHERE $u(\mathbf{x})$ IS THE SPATIAL DEPENDENT FUNCTION BY APPLYING THE MOMENTUM OPERATOR.

$$i\hbar c \nabla u(\mathbf{x}) = Eu(\mathbf{x}) \tag{4.35}$$

THE SCALE OF UNCERTAINTY IN SPACE, ϵ, ENTERS THE SPATIAL EQUATION AS A RANDOM OFFSET THAT IS GREATER THAN ZERO. IN THE SPATIAL DIFFERENTIAL EQUATION BECOMES A SECOND ORDER DIFFERENTIAL EQUATION.

$$u(\mathbf{x}) \rightarrow u(\mathbf{x}+\epsilon)$$
$$u(\mathbf{x}+\epsilon) = u(\mathbf{x}) + \epsilon u'(\mathbf{x}) \tag{4.36}$$
$$\nabla u(\mathbf{x}+\epsilon) = \nabla u(\mathbf{x}) + \epsilon \Delta u(\mathbf{x})$$

$$\{\nabla u(\mathbf{x})\}_r \rightarrow \frac{\partial u(r)}{\partial r} = u'(r) \qquad (4.37)$$

$$\{\Delta u(\mathbf{x})\}_r = \frac{\partial^2 u(r)}{\partial r^2} + \frac{n-1}{r}\frac{\partial u(r)}{\partial r} = u''(r) + \frac{n-1}{r}u'(r) \qquad (4.38)$$

THE RESULT OF EXPANDING THE DIFFERENTIAL FORMS OF THE DISPERSION RELATION WITH THE DISORDER PARAMETERS IS A PAIR OF DIFFERENTIAL EQUATION FOR THE SPATIAL VARIABLE, r, THE RADIAL COORDINATE AND THE TIME COORDINATE, τ. ACCESS TO THE ANGULAR COORDINATES IN SPHERICAL GEOMETRY ARE LOST IN THE RANDOM BEHAVIOR INTRODUCED TO GENERATE A PARTICLE DESCRIPTION LOCATED ON THE INSTANTANEOUS CENTER OF SYMMETRY OF THE PARTICLE.

USING THE COMPTON SCATTERING PARTICLE SCALE PARAMETER FOR ϵ GIVES IT A VALUE OF $\epsilon = \hbar/m_o c$. THE FIELD EQUATION ARE WRITTEN IN TERMS OF THE DIMENSION OF SPACE, n, WITH THE PARAMETERS $\gamma = E/m_o c^2$, $\omega_c = m_o c^2/\hbar$, AND $\kappa = 1/\epsilon$. THE RESULTING SPATIAL DIFFERENTIAL EQUATION FROM EXPANDING THE CONSERVATION OF ENERGY RELATION AND REFERENCED TO THE PARTICLES INSTANTANEOUS CENTER OF SYMMETRY.

$$\frac{\partial^2 u(r)}{\partial r^2} + \left(\frac{n-1}{r} + \kappa\{1 - i\gamma\}\right)\frac{\partial u(r)}{\partial r} - i\kappa^2 \gamma u(r) = 0 \qquad (4.39)$$

THE TIME DEPENDENT EQUATION CAN ALSO BE EXPANDED FROM THE THE DISPERSION RELATION $E = \hbar\omega$ WITH THE USE OF THE ENERGY OPERATOR FOR A MASSLESS FIELD.

$$\frac{\partial^2 g(\tau)}{\partial \tau^2} + (\omega_c + i\omega)\frac{\partial g(\tau)}{\partial \tau} + i\omega_c \omega g(\tau) = 0 \qquad (4.40)$$

THE SECOND ORDER SPATIAL EQUATION HAS TWO SOLUTIONS INCLUDE THE CONFLUENT HYPERGEOMETRIC FUNCTIONS $_1F_1$ AND U WHERE A AND B ARE CONSTANTS [73]:

$$u(r)_{fermion} = Ae^{-\kappa r}{}_1F_1\left[\frac{n-1}{1+i\gamma}, n-1, (1+i\gamma)\kappa r\right] \qquad (4.41)$$

$$u(r)_{boson} = Be^{-\kappa r}U[\frac{n-1}{1+i\gamma}, n-1, (1+i\gamma)\kappa r] \qquad (4.42)$$

WHAT WAS DISCOVERED ON INSPECTING THESE TWO SOLUTIONS WERE PROPERTIES CONSISTENT WITH THE TWO FAMILIES OF PARTICLES WITH MASS: BOSON AND FERMION [16]. THE FIRST SOLUTION REPRESENTS A FERMION AND THE SECOND SOLUTION REPRESENTS A BOSON BOTH WITH A REAL MASS. ALL DENSITIES DETERMINED FROM THE SOLUTIONS RETAIN SPHERICAL GROUP SYMMETRY, $U(1)$.

IN THREE DIMENSIONS THE FUNCTION $u^*(r)u(r)$ IS USED TO DIRECTLY DEFINE THE PARTICLE'S STATIC FIELD. THIS FIELD DEFINES THE PARTICLE'S STRUCTURE. CHARGE DENSITY CAN THEN BE COMPUTED FROM GAUSS'S LAW IF PARTICLE CAN SUPPORT A CHARGE [35]. THIS GENERATES THE CHARGE RADIUS OF THE PARTICLE. THE CHARGE QUANTIZATION FOR FERMIONS CAN BE DETERMINED BY AN ANALYSIS OF THE DERIVATIVE $\partial\gamma/\partial\theta$. FOR THE MASSIVE FERMION THIS WILL PRODUCE A QUANTIZED CHARGE, MASS INDEPENDENCE OF CHARGE, AND THE DIMENSIONAL DEPENDENCE OF CHARGE WHERE θ IS THE ARGUMENT OF $u(r)$ ON THE COMPLEX PLANE:

$$\theta = ArcTan\{Im[u(r)]/Re[u(r)]\}$$

IF THERE IS NO θ DEPENDENCE IN $u(r)$ THE PARTICLE HAS A ZERO CHARGE AND CANNOT SUPPORT AN ELECTROMAGNETIC TRANSITION [16].

THE SELF-REFERENCE FRAME ALSO ALLOW THE ANTI-PARTICLES TO BE DESCRIBED AS γ CAN TAKE ON A NEGATIVE VALUE [16]. THE ANTI-PARTICLE FUNCTION, $\bar{u}$, HAVE THE OPPOSITE ROTATIONAL SYMMETRY, RIGHT AND LEFT HANDED SPIRALS ON THE COMPLEX PLANE DIFFERENTIATE TO TWO TYPES. IN ADDITION IT WAS FOUND THAT $|\gamma|$ CAN TAKE ON VALUES LESS THAN ONE WHEN THE PARTICLE IS IN A BOUND STATE [13]. THESE ARE EXTENSIONS TO SPECIAL RELATIVITY THAT ARE ESSENTIAL IN DEVELOPING MATERIAL PROPERTIES.

THE MASSIVE BOSON SOLUTION IN THREE DIMENSION, $n = 3$, SHOWS A RELATIVE ENERGY DEPENDENCE THROUGH γ OF THE VALUE OF THE $u(r)$ AT THE ORIGIN THAT IS NOT FIXED AS IT IS FOR THE FERMION. THAT IS THE SOURCE OF CP VIOLATION EXPECTED FOR the massive 3D boson. The massive fermion solution in three dimensions has a fixed value of $u(r)$ at the origin with no CP problems. This has been extended to the analysis for baryons, proton and neutron, for which no CP problem were found and no need for the Axion [22].

This space, the self-reference frame, is a primitive domain where no form of momentum is defined and the dynamics only refer to the relative stability of the particles. The equations are compatible with relativity through γ, which describes their behavior with different relative observers. Linear momentum, angular momentum, spin, and the magnetic moments are dynamic properties of the laboratory frame and are not part of the particle's information developed in the self-reference frame. These properties are easily developed in the laboratory frame from the particle's structure [22].

The importance of the self-reference frame is that as a statistically independent space it can generate the particle's self-energy. Independence means there is no mapping between the two frames, either in space or time. This independence is reflected in the Pythagorean sum required for the two components in the conservation of energy relation, which adds the square of the kinetic energy to the square of the self-energy. Rather than add physical dimensions to the 3+1 space of the laboratory frame for additional particles it is possible for any particle or collection of related particles to establish an embedded private space statistically independent from the laboratory frame and from other particles and fields. This forms the basis of true superposition with no extra assumptions.

Dirac in 1932 tried to reverse his course [28] by introducing a second order field equation and private time. That ran into severe opposition from Pauli and Wessikopf [74] used a counter argument that involved the Klein-Gordon equation that does not conserve energy. Not much changed 26 years later in 1958 when Dirac who had come out with the 4th edition of the text *Quantum Mechanics* he was still

TROUBLED, AND QUOTING FROM THE LAST PARAGRAPH OF HIS BOOK, "THE DIFFICULTIES BEING OF A PROFOUND CHARACTER, CAN ONLY BE REMOVED ONLY BY SOME DRASTIC CHANGE IN THE FOUNDATION OF THE THEORY," [5]. THE DIFFICULTIES HAD TO DO WITH A SERIES OF INFINITIES GENERATED BY THE POINT PARTICLE DESCRIPTION AND HIS SEA OF NEGATIVE ENERGY POSITRONS, NEITHER OF WHICH ARE PHYSICALLY REALISTIC.

Relativistic Laboratory Frame Equation

BECAUSE OF THESE ERRORS THE ORIGINAL DIRAC EQUATION WAS ONLY PARTIALLY RELATIVISTIC AND THIS SHOWED UP IN THE SINGULARITY AT THE ORIGIN OF THE $1S$ STATE OF HYDROGEN AT THE ORIGIN [64]. THE PROBLEM IS MORE APPARENT IN COMPUTING THE GROUND STATE ENERGY OF HIGH Z ONE ELECTRON IONS. THE VALUES FROM THE DIRAC GROUND STATE ENERGY CLOSELY TRACT THOSE OF THE SCHRÖDINGER EQUATIONS [13], WHEREAS, THE BEHAVIOR FOR THE FULL RELATIVISTIC GROUND STATE ENERGY IS VERY DIFFERENT. A RELATIVISTIC ENERGY OPERATOR WHEN APPLIED IN THE LABORATORY FRAME NOT ONLY HAS TO SUPPORT DYNAMICS IT ALSO MUST INCLUDE THE PARTICLES SELF-ENERGY. THE RELATIVISTIC ENERGY OPERATOR, WHICH IS A FIRST ORDER TIME DERIVATIVE PLUS THE PARTICLE'S REST SELF-ENERGY.

$$non-relativistic \quad E \rightarrow i\hbar\frac{\partial}{\partial t}$$

$$(4.43)$$

$$relativistic \quad E \rightarrow i\hbar\frac{\partial}{\partial t} + m_o c^2$$

THE SECOND HALF OF THE DERIVATION TO COMPLIMENT THE SELF-REFERENCE FRAME PROPERTIES REQUIRES GENERATING THE COMPATIBLE DYNAMICS IN THE LABORATORY FRAME. THIS AUTOMATICALLY PRODUCES THE MECHANISM THAT GENERATES THE BASIC STATISTICAL PROPERTIES OF QUANTUM MECHANICS REQUIRED BY THE SELF-REFERENCE FRAME. TO DO THIS THE CONCEPT OF A POTENTIAL IS NECESSARY AND NOW IT IS BASED ON THE STRUCTURE OF THE PARTICLE ITSELF AS DERIVED IN THE SELF-REFERENCE FRAME. WITHIN THE

RELATIVISTIC CONSERVATION RELATION THE POTENTIAL IS DERIVED
FROM THE MASS OF THE PARTICLE. THE VARIATION $m - m_0 = \delta m_0$
REPRESENTS THE ENERGY SOURCE OF THE POTENTIAL.

$$E^2 = p^2 c^2 + (m_0 + \delta m_0)^2 c^4 \qquad (4.44)$$

$$E^2 - (m_0 c^2)^2 = p^2 c^2 + (2\delta m_0\, m_0 + \delta m_0^2)c^4 \qquad (4.45)$$

δm_0^2 IS SMALL RELATIVE TO $2\delta m_0\, m_0$ AND IS NOT DROPPED. THE
POTENTIAL IS TAKEN TO BE $V = \delta m_0 c^2$ PRODUCING AN EXACT RELA-
TIVISTIC EXPRESSION CONTAINING THE POTENTIAL.

$$E^2 - (m_0 c^2)^2 = p^2 c^2 + 2V m_0 c^2 + V^2 \qquad (4.46)$$

$$\frac{E^2 - (m_0 c^2)^2}{2m_0 c^2} = \frac{p^2}{2m_0} + V\left(1 + \frac{V}{2m_0 c^2}\right)$$

$$\frac{(E - m_0 c^2)(E + m_0 c^2)}{2m_0 c^2} = \frac{p^2}{2m_0} + V\left(1 + \frac{V}{2m_0 c^2}\right) \qquad (4.47)$$

$$\frac{i\hbar \frac{\partial}{\partial t}\left(i\hbar \frac{\partial}{\partial t} + 2m_0 c^2\right)}{2m_0 c^2} = \frac{p^2}{2m_0} + V\left(1 + \frac{V}{2m_0 c^2}\right)$$

USING THE MOMENTUM OPERATOR AND THE CORRECT ENERGY OPER-
ATOR THE EQUATION IS CONVERTED INTO THE RESULTING DIFFEREN-
TIAL EQUATION, WHICH HAS TWO ADDITIONAL TERMS ABSENT FROM
THE SCHRÖDINGER EQUATION. THE SECOND ORDER TIME DEPEN-
DENT TERM EMBEDDED THE PROPAGATING FIELD EQUATION MORE
COMMONLY FOUND FROM ELECTROMAGNETIC THEORY OF MAXWELL.
THE SECOND ADDITION IS A QUADRATIC TERM IN THE POTENTIAL,
WHOSE PRESENCE BRINGS IN THE MECHANICS OF PAIR-PRODUCTION
NATURALLY [17].

$$\frac{\hbar^2}{2m}\left\{\nabla^2 \phi - \frac{1}{c^2}\frac{\partial^2 \phi}{\partial t^2}\right\} + i\hbar \frac{\partial \phi}{\partial t} = \left(V + \frac{V^2}{2mc^2}\right)\phi \qquad (4.48)$$

THE ABOVE EQUATION REDUCED TO THE STANDARD SCHRÖDINGER EQUATION FOR SOME BOUND STATE AND FREE PROPAGATION PROBLEMS. THIS COMES AT A COST OF LOSING ITS COMPATIBILITY WITH RELATIVITY AND THE LOSS OF THE FREE FIELD EMBEDDED WAVE EQUATION. THAT REDUCTION INTRODUCES A NUMBER OF ERRORS HISTORICALLY ATTACKED BY PERTURBATION TECHNIQUES.

$V + \dfrac{V^2}{2mc^2} = 0$ Pair Production Source

THE LABORATORY FRAME DESCRIPTION OF A MASSIVE PARTICLE YIELDS THE MECHANISMS OF RANDOM BEHAVIOR NECESSARY TO PRODUCE INERTIA, IN THE QUADRATIC TERM OF THE POTENTIAL. WHEN THE POTENTIAL CONTRIBUTION IN FREE SPACE WITH NO EXTERNAL POTENTIALS THERE REMAIN TWO SOLUTIONS TO THE ABOVE EQUATION $V = 0$ AND $V = -2mc^2$, BOTH SOLUTIONS ARE EQUALLY WEIGHTED. THE SECOND SOLUTION REPRESENT A PAIR PRODUCTION ALLOWED BY THE HEISENBERG RELATION OF PARTICLE AND ANTI-PARTICLE THE SAME AS THE ORIGINAL BEING DESCRIBED. THE ANNIHILATION WITH EITHER THE ORIGINAL OR THE GENERATED PARTICLE PRODUCES THE STATISTICAL BASIS OF QUANTUM MECHANICS.

IN THE LATE 1920s MATRIX MECHANICS, SCHRÖDINGER, DIRAC, AND KLEIN-GORDON EQUATION WERE ALL ESSENTIALLY WRITTEN DOWN. THEY WERE NOT DERIVED FROM AN UNDERSTANDING OF THE CONNECTION BETWEEN RELATIVITY AND QUANTUM MECHANICS. THE DIRAC EQUATION WAS FORCED TO BE A LINEAR APPROXIMATION. THE PROBLEM THEY ALL SUFFERED FROM WAS THEY DID NOT INCLUDE THE CORRECT RELATIVISTIC BASIS. THERE IS NO SUCH THING AS A CORRECT NON-RELATIVISTIC QUANTUM DESCRIPTION, AT BEST IT IS AN APPROXIMATION THAT BARS ANY UNDERSTANDING OF PARTICLE AND FIELD STRUCTURE. THIS COLLECTION OF MISSTEPS STALLED THEORETICAL PHYSICS FOR THE NEXT $90+$ YEARS YIELDING A NUMBER OF COMPLEX WORK AROUNDS THAT YIELDED LITTLE UTILITY.

Inertia

WHAT IS REQUIRED TO GENERATE A MASS FROM A PRIMITIVE FIELD ARE OBSTACLES TO AID IN LOCALIZING A FIELD MOVING AT THE SPEED

OF LIGHT. A SET OF OBSTACLES THAT CONSERVE ENERGY IN THE LAB-
ORATORY FRAME ARE COMPOSED OF FIELD-ANTI-FIELD PAIRS. SOME-
TIME THE ORIGINAL FIELD MAKES IT THROUGH AND OTHER TIMES
IT ANNIHILATES AND ITS OPPOSITE NUMBER TAKES OVER BEING THE
PROPAGATING FIELD. THIS RESULTS IN A RANDOM DISPLACEMENT.
IF THIS PROCESS IS TRULY RANDOM THEN THE ORIGINAL FIELD WILL
BE LOCALIZED UNDER SOME VERY SPECIFIC CONDITIONS. OUR ORIG-
INAL FIELD'S A SELF-ENERGY IS TAKEN AS $\hbar\omega$ AS WELL AS FOR OUR
FINAL FIELD AS ENERGY IS CONSERVED. TO COMPUTE THE RATE OF
PAIR PRODUCTION THE SELF-ENERGY OF THE NEW PAIR BECOMES $2\hbar\omega$
WITH A MASS EQUIVALENT EQUAL TO $2m_0c^2$. THE LOCALIZATION IS
INITIATED IN THE LABORATORY FRAME SO THAT THE RATE, R, OF THE
PAIR-PRODUCTION CAN BE COMPUTED FROM THE HEISENBERG UN-
CERTAINTY RELATION FOR ENERGY.

$$R = \frac{1}{\delta t} \leq \frac{4m_0c^2}{\hbar} \tag{4.49}$$

AT ANY TIME OUR FIELD HAS A 50% CHANCE OF ENCOUNTER-
ING A PAIR AND COMPOUNDING THAT A 50% CHANCE OF ANNIHILAT-
ING AND PASSING THE BATON TO THE NEWLY MINTED FIELD. THIS
EQUAL WEIGHTING CAN BE EXPLICITLY DERIVED, SEE CHAPTER 3 IN
[17]. SO IN TOTAL IT HAS A 25% CHANCE OF BEING REPLACED. THIS
RATE TURNS INTO AN EQUALITY SINCE THE ONLY VIRTUAL FIELD PAIRS
THAT CAN INTERACT WITH ORIGINAL FIELD MUST HAVE THE IDENTI-
CAL ENERGY AS THESE ARE CONSERVATIVE PROCESSES. THIS RATE OF
REPLACEMENT IS $1/4$ THE RATE OF PAIR PRODUCTION.

$$\frac{R}{4} = \frac{m_0c^2}{\hbar} \tag{4.50}$$

THE INVERSES OF THE RATE $R/4$ IS A MEAN INTERVAL ANY PAR-
TICULAR FIELD LIVES AND THE DISTANCE LIGHT CAN TRAVEL IN THAT
INTERVAL IS ϵ WHICH NOW CAN BE COMPUTED FROM EQ. 4.50.

$$\epsilon = \frac{\hbar}{m_0c} \tag{4.51}$$

THIS IS THE COMPTON RELATION PRODUCED FROM A REAL DIS-
ORDER PARAMETER, ϵ. THE NET EFFECT ON OUR FIELD IS SET BY THE

MEAN RATE OF EXCHANGING FIELDS AND GENERATING A LOCALITY FOR A PARTICLE WITH INERTIA AS ITS LOCAL POSITION IS UNKNOWN TO A MEAN RANDOM VALUE ϵ. THE ANGULAR COORDINATE DESCRIPTION IS LOST IN THE SELF-REFERENCE FRAME AS IT IS RESET TO THE PRESENT POSITION OF THE PARTICLE'S CENTER OF SYMMETRY. BY RANDOMIZING THE LOCAL LOCATION OF THE FIELDS CENTER OF SYMMETRY A PARTICLE IS CREATED WITH A FINITE SCALE ALONG WITH LOCAL ISOTROPY. THE ORIGIN OF THE FIELD ALWAYS HAS TO KEEP SHIFTING AFTER EACH ANNIHILATION TO THE REPLACEMENT FIELD'S PARTNER. THIS RANDOM-ANNIHILATION-WALK GENERATES A LOCATION, A FUZZY LOCATION, BUT A LOCATION THAT CAN BE DESCRIBED. THE COORDINATES IN TIME AND SPACE ARE NOW STATISTICALLY INDEPENDENT OF THE ORIGINAL LABORATORY FRAME FROM WHERE THEY WERE CREATED. SO FROM THE LABORATORY FRAME WITH THE PHYSICAL PROPERTY THAT ALLOWS PAIR-PRODUCTION FOR SHORT INTERVALS A LOCALIZED ENTITY CAN BE CREATED FROM SOMETHING VERY RARE, AN ABSOLUTELY FAIR GAME OF CHANCE. THIS GAME OF CHANCE GENERATES A STATISTICALLY ISOLATED SPACE INDEPENDENT OF THE LABORATORY FRAME WITH THE PARTICLE'S INSTANTANEOUS FRAME OF REFERENCE TIED TO THE CURRENT FIELD.

Massless Fields in Self-Reference Frame

TO GENERATE A MASSLESS FIELD IN THE SELF-REFERENCE FRAME THERE ARE TWO CHOICES, EITHER SET THE MASS TO ZERO OR MAKE IT COMPLEX. THERE IS NO CHOICE IN THE LABORATORY FRAME WHERE THE MASS IS ZERO FOR A MASSLESS FIELD. IN THE SELF-REFERENCE FRAME SETTING THE MASS TO ZERO WILL NOT YIELD A STATE EQUATION FOR THE FIELD. HOWEVER, MAKING THE MASS COMPLEX IN THE SELF-REFERENCE FRAME WILL GENERATE PHYSICAL FIELD SOLUTIONS AND QUANTIZE THE FIELDS. THIS QUESTION HAS A LONG HISTORY THAT HAS PRODUCED A NUMBER OF THEORIES [75], HOWEVER, NOT UNTIL THE SELF-REFERENCE FRAME WAS FOUND DID THIS TRANSFORMATION MAKE SENSE.

MASS IS INVERSELY RELATED TO THE RANDOM VARIABLE ϵ TO MAKE MASS COMPLEX ϵ MUST BE MADE COMPLEX. BY MAKING ϵ COMPLEX IT IS EQUIVALENT TO INTRODUCING A PHASE SHIFT AND

THIS SHOULD BE RETARDED SO THE TRANSFORMATION THAT WILL BE USED IS FOUND IN EQ. 4.52 BECAUSE $\epsilon > 0$ FOR GENERATING A REAL MASS. THIS RANDOM DISPLACEMENT IS ALWAYS POSITIVE IN A SPHERICAL COORDINATE SYSTEM AS IT IS REFERENCED FROM THE INSTANTANEOUS CENTER OF SYMMETRY THAT IS CHANGING. THEREFORE FOR THE COMPLEX DISPLACEMENT THE RELATION IN EQ. 4.52 IS USED.

$$\epsilon \rightarrow -i\epsilon \tag{4.52}$$

TO TRANSFORM THE REMAINING PARAMETERS INTO FIELD EQUATIONS TO TEST THE CONJECTURE ABOUT A COMPLEX MASS, IT IS FIRST NECESSARY TO UNDERSTAND HOW γ IN THE SELF-REFERENCE FRAME TRANSFORMS.

$$\epsilon \rightarrow -i\epsilon \quad then \quad \gamma = \frac{E}{m_0 c^2} = \frac{\hbar \omega_c}{\frac{\hbar}{-i\epsilon c} c^2} = -i\frac{\epsilon}{\epsilon} = -i \tag{4.53}$$

$$\epsilon \rightarrow -i\epsilon \quad then \quad \omega_c \rightarrow i\omega_c \tag{4.54}$$

FOR THE CASE IN THE SELF-REFERENCE FRAME WHEN THE COMPTON WAVE LENGTH IS SET EQUAL TO THE RANDOM DISPLACEMENT PARAMETER, $-i\epsilon$, THEN $\gamma \rightarrow -i$. THIS IS ONE OF THE MORE IMPORTANT RELATIONSHIPS DERIVED, BECAUSE IT ESSENTIALLY ENFORCES THE QUANTIZED CONDITION ON THE RESULTANT FIELD. IN PARTICULAR THIS IS ALSO THE QUANTUM CONDITION FOR THE PHOTON ENERGY.

A PARTICLE IN THE SELF-REFERENCE FRAME TO PARTICIPATE IN AN ELECTROMAGNETIC TRANSITION OR THE EXCHANGE OF ENERGY WITH AN ELECTROSTATIC FIELD MUST BE ABLE TO CHANGE γ. FOR A MASSLESS FIELD EITHER BOSON OR FERMION IT IS NECESSARY THAT γ IS A FIXED COMPLEX CONSTANT THAT CANNOT VARY. THEREFORE, THE FIELD EITHER EXISTS OR DOESN'T EXIST WITH NO DECAY MECHANISM. THE CONSTRAINT THAT $\gamma = -i$ CONFIRMS THE ORIGINAL CONJECTURES BY PLANCK AND EINSTEIN THAT RADIATION IS QUANTIZED. THIS IS NOT THE MECHANISM OF ENERGY EXCHANGE FOR AN ELECTROSTATIC INTERACTION ONLY FOR A RADIATIVE TRANSITION FOR A REAL PHOTON.

THE SELF-REFERENCE FRAME PLACES A STRICT CONDITIONS ON THE MATERIAL PARAMETERS THAT ARE DEFINED IN THIS INDEPENDENT SPACE. IF THE EQUIVALENT COMPLEX RANDOM DISPLACEMENT

IS APPLIED TO THE PARTICLE DESCRIPTION $\epsilon \rightarrow -i\epsilon$, $\kappa \rightarrow i\kappa$ AND $\gamma \rightarrow -i$. THE MASSLESS FIELD'S DIFFERENTIAL EQUATIONS BECOME:

$$\frac{\partial^2 u(r)}{\partial r^2} + (\frac{n-1}{r})\frac{\partial u(r)}{\partial r} + \kappa^2 \gamma u(r) = 0 \qquad (4.55)$$

$$\frac{\partial^2 g(\tau)}{\partial \tau^2} + \omega^2 g(\tau) = 0 \qquad (4.56)$$

THE SOLUTIONS IN THREE DIMENSIONS ARE:

$$u_{boson} = A\, e^{-i\kappa r}\, U[1, 2, 2i\kappa r] \qquad (4.57)$$

$$u_{fermion}(r) = B\, e^{-i\kappa r}\, {}_1F_1[1, 2, 2i\kappa r] \qquad (4.58)$$

$$g(\tau) = A e^{-i\omega\tau} \qquad (4.59)$$

THE COMPLETE SOLUTIONS ARE THEN:

$$\phi(r, \tau)_{boson} = A\, e^{-i(\kappa r - \omega\tau)}\, U[1, 2, 2i\kappa r] \qquad (4.60)$$

$$\phi(r, \tau)_{fermion}(r) = B\, e^{-i(\kappa r - \omega\tau)}\, {}_1F_1[1, 2, 2i\kappa r] \qquad (4.61)$$

NOW THAT BOTH ELEMENTARY PARTICLE AND FIELD STRUCTURES HAVE BEEN DERIVED THEIR DENSITY FUNCTIONS IN THREE DIMENSIONS ARE PLOTTED IN FIGURE 4.2.

THE TOTAL WAVE FUNCTION IN THE SELF-REFERENCE FRAME $\phi(r, \tau) = u(r)g(\tau)$ THE TIME DEPENDENCE BEING OF THE FORM $e^{-i\omega\tau}$ BECOMES A CONSTANT FACTOR IN THE PROBABILITY DENSITY FUNCTION. THE PARTICLE DENSITY IN THE SELF-REFERENCE FRAME IN THREE DIMENSIONS IS GIVEN BY THE EXPRESSION $u^*(r)u(r)r^2$. THE CORE OF DENSITY $u^*(r)u(r)$ IN THE CASE OF A MASSIVE FERMION IS PROPORTIONAL TO THE STATIC ELECTRIC FIELD AND REMOVES THE $1/r^2$ SINGULARITY OF THE POINT ELECTRON AT ITS CENTER OF SYMMETRY [35]. IN THE CASE OF THE MASSIVE BOSON THE PROPERTIES OF WEAK CHARGE RESULT AND THE DESCRIPTION IS FOUND IN [16]. FOR THE MASSLESS FIELDS THE BOSON DENSITY IS A CONSTANT AS IT IS FOR THE PHOTON

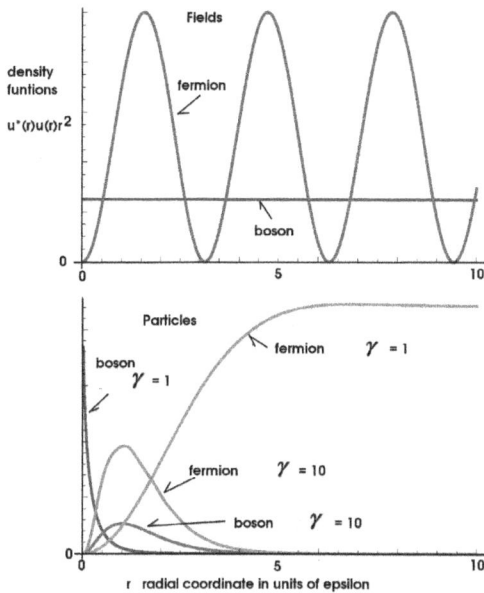

Figure 4.2: **Density functions of fields and particles in the self-reference frame. The individual density scales are arbitrary so the functions separate. The parity problem of the massive bosons can be seen at $r = 0$ as the density function in its dependence on γ and is independent of γ for fermions.**

FIELD. HOWEVER, FOR THE FERMION FIELD IT HAS A SPATIAL OSCIL-
LATORY BEHAVIOR, THAT WILL AFFECT A NUMBER OF PROPERTIES. IT
IS THE ENERGY DEPENDENT OSCILLATORY CHARACTER OF THE DEN-
SITY FUNCTION THAT IS OF PRIMARY INTEREST AS IT REDUCES THE
PARTICLES INTERACTION CROSS SECTION.

$$boson$$

$$< u^*_{photon} u_{photon} r^2 > \; = \; 1$$

$$(4.62)$$

$$fermion$$

$$< u^*_{neutrino}(r) u_{neutrino}(r) r^2 > \; = \; < Sin^2 \kappa r > \; = \; \frac{1}{2}$$

THE MEAN VALUE OF THE Sin^2 TERM IS EXACTLY ONE HALF. THIS BE-

HAVIOR IN THE SPATIAL PORTION OF THE WAVE FUNCTION IS UNIQUE AMONG PARTICLES AND WILL LEAD TO A REDUCTION IN DETECTED SENSITIVITY BY EXACTLY 50% IN MEASURED DATA WHETHER FROM SOLAR OR REACTOR GENERATED ELECTRON NEUTRINOS.

THE EXTENSIVE LITERATURE ON THE NEUTRINO-CROSS SECTION AS A FUNCTION OF ENERGY THAT RESULT ARE DYNAMIC CALCULATIONS AT A LEVEL ABOVE OF THE DENSITY CALCULATION FOR THE NEUTRINO IN THE SELF-REFERENCE FRAME [76]. THE KINEMATIC MODELS DO NOT INVOLVE THE STRUCTURE OF THE PARTICLES THEMSELVES, ONLY THEIR BULK PROPERTIES AND ALLOWED INTERACTIONS. IT IS NOT NECESSARY TO INVOLVE THE SPECIFIC MECHANISMS FOR THE ENERGY DEPENDENT CALCULATION OF CROSS-SECTIONS, BECAUSE THE CORRECTION BEING INTRODUCED WILL AFFECT THE NEUTRINO ACROSS ITS ENTIRE ENERGY RANGE UNIFORMLY.

IN THREE DIMENSIONS THE COMPLEX MASS GENERATES TWO MASSLESS FIELDS THAT APPEAR TO HAVE REAL PHYSICAL COUNTER PARTS: PHOTON AND THE ELECTRON NEUTRINO. FIRST IS A BOSON WITH A UNIT DENSITY CHARACTERISTIC OF A BASIC PHOTON AND THEN A MASSLESS ELEMENTARY FERMION REPRESENTING A NEUTRINO. THESE ARE SOLUTIONS IN THE SELF-REFERENCE FRAME AND NOT IN THE LABORATORY FRAME WHERE THEIR COMPLETE STRUCTURE IS DEVELOPED. BOTH SOLUTIONS ARE OF MASSLESS FIELDS SHOWING NO PREFERRED LOCAL STRUCTURE. THIS WAS FORCED BY $\gamma = -i$ BEING FIXED COMPLEX CONSTANT. ANY OTHER VALUES OF γ PRODUCED DIVERGENT SOLUTIONS THAT ARE NOT VALID. DIVERGENCE HERE MEANS THAT THE DENSITY FUNCTIONS GROW LARGER WITH INCREASING R, WHICH IS NEITHER THE PROPERTY OF A PHYSICAL REALIZABLE PARTICLE OR FIELD. FIXING γ FOR MASSLESS FIELD ALSO INSURES THE INDEPENDENCE OF THE SPEED OF LIGHT IN ANY REFERENCE FRAME. THIS RESTRICTION ON γ IS A REQUIREMENT FOR THE QUANTIZATION OF THE FIELD FOR BOTH THE PHOTON AND NEUTRINO.

IN THE SELF-REFERENCE FRAME THE HARMONIC TIME DEPENDENCE OF A STABLE ENTITY THAT STARTS WITH A PRIVATE TIME DEPENDENCE WHEN THE FRAME IS CREATED WITH NO PREVIOUS HISTORY. ALL ENTITIES WHETHER A PARTICLE OR A FIELD COME WITH THEIR OWN CLOCKS, VIA THEIR TIME DEPENDENCE, AND ARE ESSENTIALLY ISOLATED BY THE STATISTICAL INDEPENDENCE OF THE SPACE

IN WHICH THEY WERE GENERATED. THE ONLY EXCEPTION IS WHEN TWO OR MORE PARTICLES SHARE THE SAME CLOCK EITHER FROM BEING CREATED AT THE SAME INSTANCE OR INTERACTING WITH ONE OF TWO FIELDS OR PARTICLES THAT WERE CREATED AS A PAIR. THIS BEHAVIOR OF PARTICLES AND FIELDS SHARING A SELF-REFERENCE FRAME IS IMPORTANT FOR UNDERSTANDING ENTANGLEMENT.

THE ORIGINAL REQUIREMENT FOR SPECIAL RELATIVITY AS LAID OUT BY EINSTEIN 1905 ARE THE EXISTENCE OF A MEASUREMENT SCALE AND A TIME BASE. BOTH CONDITIONS ARE SATISFIED FOR EACH INDIVIDUAL PARTICLE AND QUANTIZED FIELD BY THEIR PROPERTIES IN THEIR SELF-REFERENCE FRAME. NO EXTERNAL OBSERVER IS REQUIRED TO FULFILL THESE NEEDS FOR A CLOCK AND A RULER.

Massless Fields in the Laboratory Frame

TAKING EQUATION 4.63 AND SETTING MASS TO ZERO YIELDS THE WAVE EQUATION FOR MASSLESS FIELDS WITH AN INTERACTION TERM THAT GENERATES THE REFRACTIVE RESPONSE WHEN THE FIELDS ENCOUNTER MATTER.

$$\nabla^2 \phi - \frac{1}{c^2}\frac{\partial^2 \phi}{\partial t^2} = \frac{V^2}{\hbar^2 c^2}\phi \tag{4.63}$$

THIS MAKES A MORE GENERAL COUPLING BETWEEN QUANTUM MECHANICS AND ELECTROMAGNETIC THEORY AS QUANTUM MECHANICS GENERATES THE PROPAGATING FIELD BEHAVIOR THAT IS FOUND IN THE FARADAY-MAXWELL THEORY. THIS EQUATION IS EQUALLY VALID FOR ANY MASSLESS PROPAGATING FIELD SUCH AS THE NEUTRINO. NOW REFRACTION FOR BOTH THE PHOTON AND NEUTRINO CAN BE TREATED AS A POTENTIAL INTERACTION WITH A PHYSICAL MEDIUM.

4.2   Binding

EXTRACTED FROM [66]

THE ROLE SPECIAL RELATIVITY PLAYS IN BOUND STRUCTURES CAN BE EXPLOITED TO DEFINE HOW LOWER DIMENSIONAL COMPONENTS

BIND TO FORM THE TWO PRINCIPAL NUCLEONS WITHOUT SHUTTLING COLORED GAUGE BOSONS. FOR THE HYDROGEN GROUND STATE THE $|\gamma|$ OF SPECIAL RELATIVITY TAKES ON A VALUE LESS THAN ONE. THE CLOSED FORM SOLUTION OF BINDING THAT SATISFIES RELATIVISTIC CONSERVATION OF ENERGY FOR THE GROUND STATE OF THE HYDROGEN ATOM REQUIRES NO CORRECTIONS FROM QUANTUM ELECTRODYNAMICS IF RELATIVITY IS PROPERLY TREATED [13]. THE ELECTRON GIVES UP SOME SELF-ENERGY, MASS, TO BIND. THE LOSS OF MASS IS MORE OBVIOUS FOR NUCLEON BINDING. FROM THE COMPTON RELATIONSHIP THE SCALE OF THE INDIVIDUAL PARTICLES, $\langle r \rangle$ INCREASES WHEN BOUND BECAUSE OF THE LOSS OF MASS. THE ONLY REQUIREMENT FOR BINDING OF TWO OR MORE COMPONENTS IS THAT THE SCALE INCREASE WITH DECREASING γ WHEN THE COMPONENTS OVERLAP WITH THE SAME CENTER OF SYMMETRY. THIS MEANS THE COMPONENTS ARE ATTRACTING EACH OTHER. IT IS NOT DIFFICULT TO COMPUTE THE SCALE WHERE $H(r)$ IS THE WAVE FUNCTION OF THE COMBINED COMPONENTS INTEGRATED OVER THE VOLUME TO GENERATE THE MEAN SCALE $\langle r \rangle$ THAT IS INVERSELY PROPORTIONAL TO MASS, SEE EQUATION 4.64.

$$\langle r \rangle (\gamma, n, m) \; = \frac{\int H^* \, r \, H dv}{\int H^* \, H dv} \tag{4.64}$$

AN EXAMPLE OF TWO DIFFERENT COMPONENTS WITH DIMENSIONS, n AND m, $H(r)$ IS CONSTRUCTED AS A PRODUCT OF THESE LOWER DIMENSIONAL COMPONENTS WITH THEIR OWN JACOBIAN SCALE AND TAKES A FORM FOUND IN EQUATION 4.65.

$$\langle r \rangle (\gamma, n, m) \; = \frac{\int u^*(r,n) u^*(r,m) r^{n+m-1} u(r,n) u(r,m) dr}{\int u^*(r,\gamma,n) u^*(r,m) r^{n+m-2} u(r,n) u(r,m) dr} \tag{4.65}$$

BINDING (*stable*) $\langle r \rangle_{|\gamma|<1} > \langle r \rangle_{|\gamma|=1}$

$$\tag{4.66}$$

No BINDING (*unstable*) $\langle r \rangle_{|\gamma|<1} < \langle r \rangle_{|\gamma|=1}$

ONLY THE SPATIAL PART OF THE WAVE FUNCTIONS WILL BE USED AS THE TIME DEPENDENT PRODUCTS PRODUCES A CONSTANT FACTOR OF 1 IN THE COMPUTATION. THE RELATIVE SCALES OF THE

INDIVIDUAL COMPONENTS WILL PLAY A MAJOR PART IN DETERMIN-
ING WHICH COMBINATIONS WILL FORM STABLE STRUCTURES. IN THE
SELF-REFERENCE FRAME WHERE THESE LOWER DIMENSIONAL COM-
PONENTS ARE ORGANIZED THERE ARE NO INDIVIDUAL DYNAMICAL
CONTRIBUTIONS AS FOUND IN THE LABORATORY FRAME FOR WHOLE
THREE DIMENSIONAL ENTITIES.

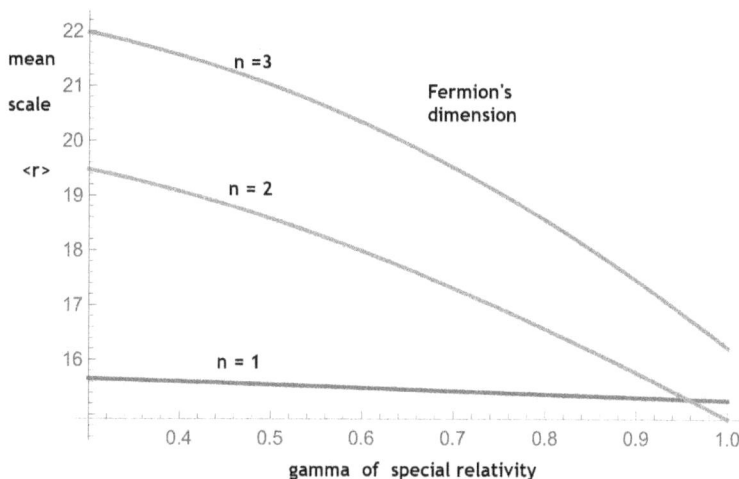

Figure 4.3: As the mean scale $\langle r \rangle$ increases the particle loses mass and the energy is used in bind to another particle, using equation 4.64.

4.3 Abstracts & URLs

Electrodynamics in Iron and Steel

In order to calculate the reflected EM fields at low amplitudes in iron and steel, more must be understood about the nature of long wavelength excitation in these metals. A bulk piece of iron is a very complex material with micro-structure, a split band structure, magnetic domains and crystallographic textures that affect domain orientation. Probing iron and other bulk ferro-magnetic materials with weak reflected and transmitted induc-

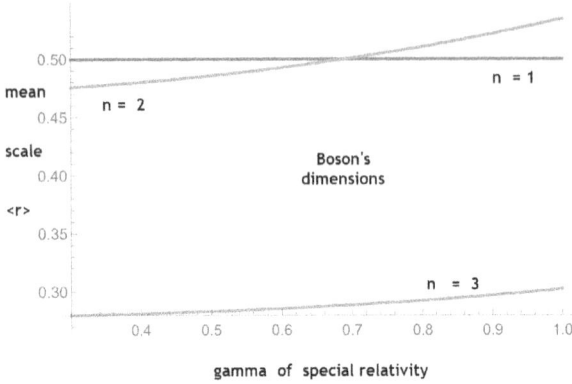

Figure 4.4: **The 3D boson, $W^{\pm}Z^{o}$, shows no scale increase with decreasing γ and therefore cannot be found in a bound structure because it will supply no energy to the bond.**

tive low frequency fields is an easy operation to perform but the responses are difficult to interpret because of the complexity and variety of the structures affected by the fields. First starting with a simple single coil induction measurement and classical EM calculation to show the error is grossly under estimating the measured response. Extending this experiment to measuring the transmission of the induced fields allows the extraction of three dispersion curves which define these internal fields. One dispersion curve yielded an exceedingly small effective mass for those spin waves, $1.3 \times 10^{-9} \ m_{electron}$. There is a second distinct dispersion curve more representative of the density function of a zero momentum bound state rather than a propagating wave. The third dispersion curve describes a magneto-elastic coupling to a very long wave length propagating mode. . . .

https://arxiv.org/pdf/0910.1631

Spintronics Enters the Iron Age

In this paper, a set of spintronics circuit elements are introduced which can be used to build complete analog circuits.

This allows circuit development using stable quantum states in the bulk of iron based alloys. As an example application a simple circuit is used to learn about hydrogen in iron whose minor concentration plays a large role in altering the activity of the Bose–Einstein-like condensations under measurement.

Journal of Metals Vol 61 No. 6 page 67 2009
https://www.castinganalysis.com/files/spintronics.pdf

Proton in SRF Niobium

Hydrogen is a difficult impurity to physically deal with in superconducting radio frequency (SRF) niobium, therefore, its properties in the metals should be well understood to allow the metal's superconducting properties to be optimized for minimum loss in the construction of resonant accelerator cavities. It is known that hydrogen is a paramagnetic impurity in niobium from NMR studies. This paramagnetism and its effect on superconducting properties are important to understand. To that end analytical induction measurements aimed at isolating the magnetic properties of hydrogen in SRF niobium are introduced along with optical reflection spectroscopy which is also sensitive to the presence of hydrogen. . . . .

AIP Conf. Proc. **1352**, pages 205-335 (2011)

Relativistic Longitudinal Spin Wave

In a study of magnetic losses in iron and steel a relativistic longitudinal spin wave was found. The exceedingly small mass and large scale of the spin wave requires an accurate relativistic description for a boson. Because of these characteristics it forms a state that is decoupled from property variations of the substrate on a small scale and is only weakly dissipated. In the effort to explain the behavior of this spin wave an elementary quantum representation of a particle was found to be provided by a differential equation which produces two solutions: one for boson family and one for fermion family of elementary particles. This derivation of a local statistical quantum state equation from the massless dispersion relation provides a general mechanism for obtaining a statistical description required for quantum mechanics and produces a definition of a quantum particle. The new analysis allowed confirmation of the mass value of the longitudinal spin wave from the original experimental measurements. The analysis required the introduction of a reference frame located with the particle which resulted in a quantum description that is consistent with the principles of relativity.

https://vixra.org/pdf/1405.0015v1.pdf
https://www.castinganalysis.com/files/rlsw.pdf

Electrostatics

Quantum mechanics should be able to generate the basic properties of a particle. One of the most basic properties is charge and the associated electrostatic electric field. Electrostatic force is a fundamental characteristics of a charged fermion and should have its nature described by the fermion's structure. To produce the particle properties require two spaces that define both their dynamics and their base structure. Relativity and the conservation of energy dictate how these two separate spaces are connected and the differential equations that describe behavior within these two spaces. The main static characteristics of an elementary fermion are mass and charge. Mass represents a scale measure of the fermion and it appears that charge results from the detailed structure of the fermion, which must merge into the electric field description of Maxwell. Coulomb's law is a good approximation for large distances, but it is a poor approximation at dimension on the order of a particle's Compton wavelength. The relativistic description of the fermion in its own frame of reference contains the information required for producing the electrostatic field over all space without a singularity as a source. . . .

AIP Conf. Vol 1697, pp 040004-1-14, AIP Melville NY 2015
https://www.castinganalysis.com/files/electrostatics.pdf

Refraction

Refraction is treated classically, which is not physically realistic. Unlike optical reflection that is well understood refraction, is a more difficult problem exposing a major missing piece of quantum mechanics. Refraction normally is treated either classically or as a non-relativistic perturbation response. Recently it became apparent where this property found its quantum origin in a full relativistic quantum description.

https://vixra.org/pdf/1809.0582v3.pdf
https://www.castinganalysis.com/files/refraction.pdf

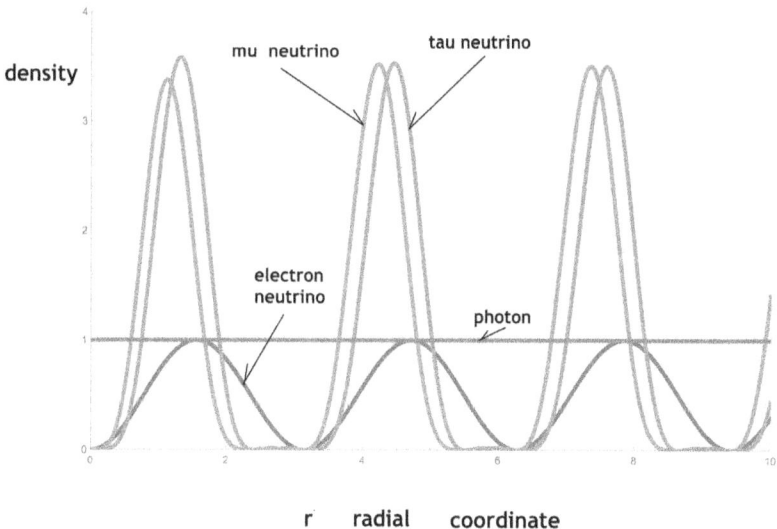

Figure 4.5: ν_μ and ν_τ neutrino normalized spatial density function are not very different from the massless ν_e electron neutrino. The electron neutrion's density function is $Sin^2(\kappa r)$. The photon's density function in the self-reference frame plot is a straight horizontal line with an amplitude of one, where $\kappa = 1$ [22]. Figure taken from [10].

2019 Christmas Card

Not-so-Standard Model of particle classification is presented in a chart.

https://www.castinganalysis.com/files/christmas_card.pdf

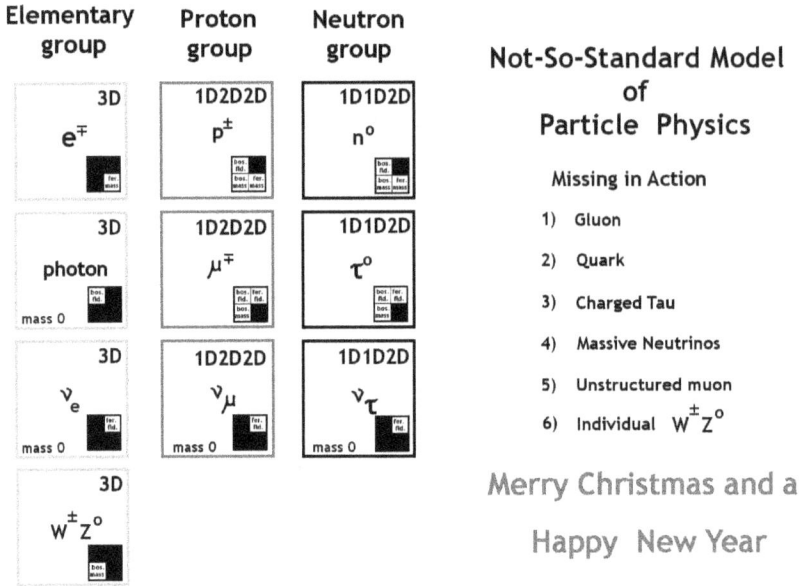

Figure 4.6: **Not So Standard Model of 2019 where particles are made from the massive and massless components in 1,2, and 3 dimensions of the boson and fermion solutions to the self-reference frame wave equations, see Section 4.4.**

Nuclear Structure and Cold Fusion

Combining advances in understanding the strong force with experiments on lattice fusion allows a description of lattice D–D fusion to be constructed. What has to be exposed is a nuclear energy loss mechanism leaving little or no residual radioactivity. The requirements on the lattice for D–D fusion are strict and appear to be limited to particular FCC lattices within a range of lattice parameters. A mechanical understanding is needed of how elevated local concentrations of deuterium are achieved while avoiding being trapped at defect sites. Using optical, RF, and experimental anomalous heat data the metallurgical requirements for the process is refined by considering a combination of diffusion, partial molar volume, positron annihilation data, defect kinetics, and electronic band structure allowing logical exhaustion to identify the kinetic structure that drives lattice nuclear fusion.

J. Condensed Matter Nucl. Sci. 30 (2020) 1-24
https://www.castinganalysis.com/files/
    nuclear_structure_and_cf.pdf

Big Bang's Quantum Problem

The early twentieth century produced the beginnings of relativity, quantum mechanics, and the big bang, but then went off the rails like much of the world in the early 1930s. The rest of the world recovered, but quantum mechanics did not recover. Physics was weighed down with a continuum geometry that did not allow quantum mechanics and relativity to be united. . . .

Infinite Energy vol 157 p 15 2021
https://vixra.org/pdf/2107.0123v3.pdf
https://www.castinganalysis.com/files/big_bang_rev3.pdf

Bound State

The ground state energy of atomic hydrogen is difficult to measure precisely. This allowed theorists to bypass the role relativity plays in setting the energy levels. Rather they concentrated on lesser effects. Data for hydrogen like ions at high Z confirmed the strong role relativity plays in bonding showing how special relativity's γ takes on values less than one in the bound state. Another major omission was found dating to the 1920s on how the asymmetrical binding fields of the proton and electron must be treated before allowing the atomic hydrogen states to be solved by the standard perturbation theory.

https://vixra.org/pdf/2103.0026v3.pdf

https://www.castinganalysis.com/files/boundstate.pdf current revision

21cm Quantum Amplifier

Hydrogen being the most common element in the universe is almost invisible in atomic form though it is common as a as minor contaminating component in most terrestrial compounds. Atomic hydrogen and its isotopes are the only chemically active atoms whose valence electron is not screened from the nucleus. This unique property leads to a rich spectroscopic behavior when weakly bonded to other molecules, surfaces, or embedded within solids. The spectra's origin lie in rotational nuclear degrees of freedom that become active when the atoms are polarization bonded to other structures. Free neutral atomic hydrogen is difficult to detect by its 1420.4 MHz emission even in objects as large as the local Virgo cluster of galaxies. Our surprise was in detecting intense signals with an inexpensive receiver near 1420.4 MHz in the spectral band reserved for radio astronomy where broadcasting is forbidden. These signals behaved like emissions from slightly perturbed 1S atomic hydrogen possessing rotational states with very small energy shifts. These signals are ubiquitous when there is any low level electromagnetic noise present.

https://vixra.org/pdf/2204.0024v3.pdf
https://www.castinganalysis.com/files/
    21cm-quantum-amplifer-v3.pdf

Statistical Independence in Quantum Mechanics

Algebraic mistakes of using a non-relativistic function betrayed Dirac's elegant derivation of the relativistic equation of quantum mechanics and exposed a short coming of special relativity. It was a serious mistake because that famous paper became a model for theorist to follow who produced an unending stream of nonsense. The mistake was compounded because it hid the fact that special relativity was still incomplete. Multiple independent spaces are required to generate both dynamics as well as produce particle properties. The concept of statistical independence of spaces that encapsulated quantum objects, fields and particles, was necessary for physics to have a relativistic basis for both massive particles and massless fields. The example that will be developed is the origin of the solar neutrino survival data that requires the electron neutrino to be massless as originally proposed by Pauli. The analysis renders a proof of the original quantum conjecture by Planck and Einstein that radiation is quantized and how inertia for massive particles is generated.

https://vixra.org/pdf/2303.0047v3.pdf

https://www.castinganalysis.com/files/
    statistical_independence_in_qm.pdf

https://academia.edu/37049106/statistical
    _independence_in_quantum_mechanics

Origin of the Meissner Effect and Superconductivity

Superconductivity in most metals is due to the activity of longitudinal spin waves binding electrons into pairs in such a way that the Meissner effect is generated along with the angular momentum responses in static magnetic fields. The bulk of these spin waves appear to be sourced by nuclear spins on the lattice. Experimentally longitudinal spin waves are not difficult to detect at room temperature as they form Bose-Einstein condensates that have onset temperatures, $> 1000^o K$, for the low mass entities, less than 10^{-40} kg. These large scale quantum structures on the order of 1 meter are ubiquitous in metals and will also exist in space with low density matter where the ambient static magnetic fields are weak and temperatures are low. These massive boson collections probably are the source of the gravitationally detected dark matter in space and these experiments provide a test bed to understand their properties.

https://vixra.org/pdf/2402.0017v2.pdf

https://www.castinganalysis.com/files/

nuclear-spin-wave-v3.pdf

Nucleons

The nucleons with their components and force fields are a greater challenge for quantum mechanics than the hydrogen atom that had provided deep cover to hide the defects both in quantum mechanics and relativity [67]. Once a correction to special relativity was made allowing γ to take on values less than one the hydrogen ground state could be accurately computed providing a method to deal with the nucleons. A result was to discover why the neutron and proton are stable and have similar masses.

https://vixra.org/pdf/2506.0155v1.pdf

https://www.castinganalysis.com/files/nucleons.pdf

Open Letter to DoE on Quantum Computing

https://www.castinganalysis.com/files/

DoE_comment_rev1.pdf

4.4 Equations

4.4.1 Laboratory Frame

Laboratory Frame: *V for Non−gravitational Potential*
from $E^2 = p^2c^2 + (m_o + \delta m)^2 c^4$ *where* $V = \delta mc^2$

Rest State Equation

$$\nabla^2\Phi - \frac{1}{c^2}\frac{\partial^2\Phi}{\partial t^2} = \frac{2m_o}{\hbar^2}\{-i\hbar\frac{\partial\Phi}{\partial t} + V(1 + \frac{V}{2m_oc^2})\Phi\}$$

Dynamic Equation

$$i\hbar\frac{\partial\Phi}{\partial t} = -\frac{\hbar^2}{m_o(1+\gamma)}\nabla^2\Phi + \frac{2V}{1+\gamma}(1 + \frac{V}{2m_oc^2})\Phi$$

Laboratory Frame: V_g *Gravitational Potential*

from $(E + V_g)^2 = p^2c^2 + (mc^2)^2$

$$\frac{\hbar}{i}\frac{\partial\Phi}{\partial t} = -\frac{\hbar^2c^2}{(1+\gamma)mc^2 + V_g}\nabla^2\Phi - V_g\Phi \qquad dynamic$$

$$\frac{\partial^2\Phi}{\partial t^2} + \frac{2i}{\hbar}(V_g + mc^2)\frac{\partial\Phi}{\partial t} - \frac{2V_gmc^2 + V_g^2}{\hbar^2}\Phi = c^2\nabla\Phi$$

For a massless field : photon or neutrino :
$$\hbar\omega_o + V_g = \hbar\omega$$

(4.67)

Table 4.1: **Correcting the Schrödinger equation.** Where $\gamma = E/m_0 c^2$ is the substitution used in the dynamic equation on the right. All equations and table taken from the 2nd and 3rd edition of *"yes Virginia" quantum mechanics can be understood* [16]

Relative Rest State	Dynamic
$\dfrac{i\hbar\frac{\partial}{\partial t}\left(i\hbar\frac{\partial}{\partial t}+2m_0 c^2\right)}{2m_0 c^2} = \dfrac{p^2}{2m_0} + V + \dfrac{V^2}{2m_0 c^2}$	$\dfrac{m_0 c^2(1+\gamma)(E-m_0 c^2)}{2m_0 c^2} = \dfrac{p^2}{2m_0} + V + \dfrac{V^2}{2m_0 c^2}$
$\downarrow$	$\downarrow$
$-\dfrac{\hbar^2}{2m_0 c^2}\dfrac{\partial^2}{\partial t^2} + i\hbar\dfrac{\partial}{\partial t} = \dfrac{p^2}{2m_0} + V + \dfrac{V^2}{2m_0 c^2}$	$i\hbar\dfrac{\partial}{\partial t} = \dfrac{p^2}{m_0(1+\gamma)} + \dfrac{2V}{1+\gamma}\left(1 + \dfrac{V}{2m_0 c^2}\right)$
$\downarrow$	$\downarrow$
for $V = 0$	for $V = 0$
$\downarrow$	$\downarrow$
$-\dfrac{\hbar^2}{2m_0 c^2}\dfrac{\partial^2}{\partial t^2} + i\hbar\dfrac{\partial}{\partial t} = \dfrac{p^2}{2m_0}$	$i\hbar\dfrac{\partial}{\partial t} = \dfrac{p^2}{m_0(1+\gamma)}$
momentum operator	momentum operator
$\downarrow$	$\downarrow$
$\nabla^2 - \dfrac{1}{c^2}\dfrac{\partial^2}{\partial t^2} + \dfrac{2m_0 i}{\hbar}\dfrac{\partial}{\partial t} = 0$	$i\hbar\dfrac{\partial}{\partial t} = -\dfrac{\hbar^2\nabla^2}{m_0(1+\gamma)}$
Relativistic wave equation	Schrödinger Equation +
particle structure	particle dynamics
$\downarrow$	$\downarrow$
$m = 0$	with
$\downarrow$	Lorentz contraction
wave equation	& time dilation

4.4.2 Massive Particle Self-Reference Frame Equations

Self-Reference Frame: *Starting Equations*

$$E = \hbar\omega \qquad E = cp \qquad n \ Dimension$$

Parmeters :

$$\kappa = \frac{1}{\epsilon} \ Propagation \ Factor \qquad \gamma = \frac{E}{m_o c^2} \quad Relativity$$

$$\omega_c = \frac{m_o c^2}{\hbar} \quad Field \ Energy$$

Massive Particle Wave Equations

$$Spatial \ Dependence: \quad \frac{\partial^2 \mathbf{u}(r)}{\partial r^2} + \left(\frac{n-1}{r} - \kappa\{1 - i\gamma\}\right)\frac{\partial \mathbf{u}(r)}{\partial r} + i\gamma\kappa^2 \mathbf{u}(r) = 0$$

$$Time \ Dependence: \qquad \frac{\partial^2 g(\tau)}{\partial \tau^2} + \{\omega_c + i\omega\}\frac{\partial g(\tau)}{\partial \tau} + i\omega\omega_c g(\tau) = 0$$

In the self-reference frame γ
can take on negative values and values less than one
spatial massive particles are in boldface $\mathbf{u}(r)$

$$fermion: \quad \mathbf{u}(r) \sim e^{-\kappa r} {}_1F_1\left[\frac{n-1}{1+i\gamma}, n-1, (1+i\gamma)\kappa r\right]; n > 1$$

$$\mathbf{u}(r) \sim \{(1+i\gamma)\kappa r\}^{2-n} e^{-\kappa r} {}_1F_1\left[\frac{n-1}{1+i\gamma} + 2 - n, 3 - n, (1+i\gamma)\kappa r\right]; \ n < 2$$

$$boson: \quad \mathbf{u}(r) \sim e^{-\kappa r} U\left[\frac{n-1}{1+i\gamma}, n-1, (1+i\gamma)\kappa r\right]$$

$$(4.68)$$

4.4.3 Massless Field Self-Reference Frame Equations

3D Solutions in Self-Reference Frame for Massless Fields

$Transformations\ takes\ the\ Massive\ Longitudinal\ Fields$

$into\ the\ Massless\ Transverse\ Fields$

$Transformations: \epsilon \rightarrow -i\epsilon\ \&\ \kappa \rightarrow i\kappa\ \&\ \gamma : 1 \rightarrow -i\ \&\ \omega_c \rightarrow i\omega_c$

$$fermion \quad u(r) \sim e^{-i\kappa r}{}_1F_1[\frac{n-1}{2}, n-1, 2i\kappa r]\ ;\ n > 1$$

$$u(r) \sim (2i\kappa r)^{2-n}e^{-i\kappa r}{}_1F_1[\frac{3-n}{2}, 3-n, 2i\kappa r]\ ;\ n < 2$$

$$boson \quad u(r) \sim e^{-i\kappa r}U[\frac{n-1}{2}, n-1, 2i\kappa r]$$

$$in\ 3\ dimensions$$

$$Photon \quad \rightarrow \quad u(r) = e^{-i\kappa r}U[1, 2, 2i\kappa r] \rightarrow u(r) = \frac{e^{-i\kappa r}}{r}$$

$$electron\ neutrino\ \nu \rightarrow u(r) = e^{-i\kappa r}{}_1F_1[1, 2, 2i\kappa r] \rightarrow u(r) = \frac{Sin(\kappa r)}{r}$$

$_1F_1[a,b,c]\ \ \&\ \ U[a,b,c]\ are\ confluent\ hypergeometric\ functions.$

$Density\ of\ the\ particle\ or\ the\ field\ is: \quad u^*(r)u(r)r^{n-1}$

$Strength\ of\ the\ particle\ field\ is: \quad u^*(r)u(r)$

$Derivations\ found\ in\ [16]\ [17]\ [22]$

$$(4.69)$$

4.4.4 Activity Map

gravitational potential

$$(E + V_g)^2 = p^2 c^2 + (m c^2)^2$$

$$\frac{\hbar}{i}\frac{\partial \Phi}{\partial t} = -\frac{\hbar^2 c^2}{(1+\gamma)m c^2 + V_g}\nabla^2\Phi - V_g\Phi \qquad dynamic$$

$$\frac{\partial^2\Phi}{\partial t^2} + \frac{2i}{\hbar}(V_g + m c^2)\frac{\partial\Phi}{\partial t} - \frac{2V_g m c^2 + V_g^2}{\hbar^2}\Phi = c^2\nabla\Phi$$

For a massless field : photon or neutrino :
$$\hbar\omega_o + V_g = \hbar\omega$$

electrostatic & strong force potential

$$E^2 = p^2 c^2 + (m_o + \delta m)^2 c^4$$

$$E^2 = p^2 c^2 + (m_o c^2)^2 + 2V m_o c^2 + V^2$$

$$\nabla^2\Phi - \frac{1}{c^2}\frac{\partial^2\Phi}{\partial t^2} = \frac{2m_o}{\hbar^2}\{-i\hbar\frac{\partial\Phi}{\partial t} + V(1 + \frac{V}{2m_o c^2})\Phi\}$$

$$i\hbar\frac{\partial\Phi}{\partial t} = -\frac{\hbar^2}{m_o(1+\gamma)}\nabla^2\Phi + \frac{2V}{1+\gamma}(1+\frac{V}{2m_o c^2})\Phi$$

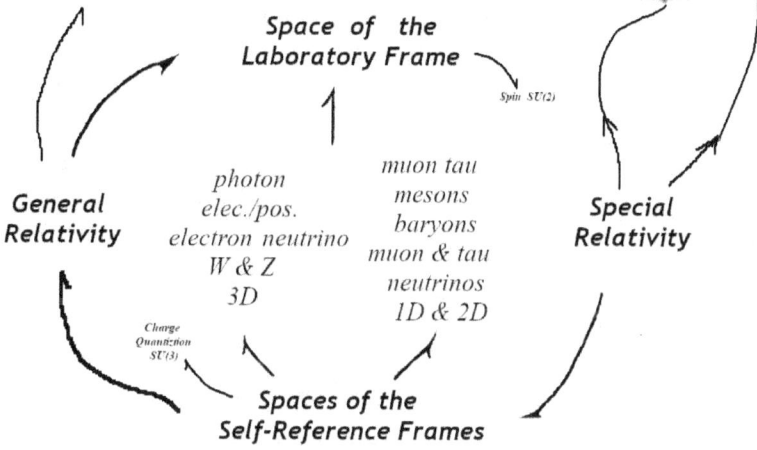

Space of the Laboratory Frame

Spin SU(2)

General Relativity

photon
elec./pos.
electron neutrino
W & Z
3D

muon tau
mesons
baryons
muon & tau
neutrinos
1D & 2D

Special Relativity

Charge Quantization SU(3)

Spaces of the Self-Reference Frames

$$\frac{\partial^2 u(r)}{\partial r^2} + (\frac{n-1}{r} - \kappa\{1-i\gamma\})\frac{\partial u(r)}{\partial r} + i\gamma\kappa^2 u(r) = 0$$

$$\frac{\partial^2 g(\tau)}{\partial \tau^2} + \{\omega_c + i\omega\}\frac{\partial g(\tau)}{\partial \tau} + i\omega\omega_c g(\tau) = 0$$

Transformations takes the Massive Longitudinal Fields
into the Massless Transverse Fields

$$c \to -ic \quad \& \quad \kappa \to i\kappa \quad \& \quad \gamma : 1 \to -i \quad \& \quad \omega_c \to i\omega_c$$

All solution U(1) symmetry
in self-reference frame

Acknowledgments

Polykarp Kusch was an excellent teacher and researcher, who measured the magnetic moment of the electron. By the early 1960s he was convinced that theoretical physics had broken down in the 1930s. His research project on an accurate determination of the hydrogen ground state in 1967 was to expose the defects in quantum electrodynamics. While being involved in that work he encouraged a reanalysis of quantum mechanics. Secondly, Jack Steinberger whose freshman course in quantum mechanics exposed two problems. First, what does a complex mass represent and secondly, why were longitudinal fields under represented in physics? These are old problems that both these gentlemen shared with the freshman. John David Jackson helped to restart this work by getting our original experimental work published that gave the first glimpse of the self-reference frame.

David Schwartz recalled that after the experiment on detecting the muon neutrino was completed, P. Kusch told his father Melvin Schwartz *"you will never do anything better, go to law school and earn some money"*. There was more to the quote than a sly joke. In 1959 Kusch had no disparaging remarks to make on quantum electrodynamics by 1967 he was working to show it was not physics. Jerry Stefani and Thomas Allen were able to go through Kusch's papers at the University of Texas library in Dallas that shed more light on his thinking.   Jerry Dunn's help with the book's production was invaluable.

Additional Reading

"I cannot live without books"

THOMAS JEFFERSON TO JOHN ADAMS 1815

James Ferguson on Technology *Lectures on Selected Topics* various editions from 1761 to 1825

James R. Newman *The World of Mathematics* 4 volumes Dover Press

Max Born *Atomic Physics* Dover Press

Enrico Fermi *Thermodynamics* Dover Press

Allan Cottrell *An Introduction to Metallurgy* St. Martins Press

Paul Nahim on complex numbers *An Imaginary Tale 2nd edition*

Silvanus P. Thompson *Calculus Made Easy*

W. Kaplan *Advanced Calculus*

Watson & Whitaker *Modern Analysis* Cambridge Univ. Press

Joseph Dauben *Georg Cantor*

John D. Jackson *Classical Electrodynamics* 2nd or 3rd editions

Leonard Schiff *Quantum Mechanics*

Paul Dirac *The Principles of Quiantum Mechanics* 4th edition

A. Einstein *The Principle of Relativity*

Bibliography

[1] F. Del Santo. The foundation of quantum mechanics in post-war italy's cultural context arxiv:2011.11969, 2020.

[2] H. MacInnes. *Neither Five nor Three*. Harcourt, Brace & World, NYC, 1951.

[3] W. Gordon. Die energienveaus des wasserstoffatoms nach der diracschen quantentheorie des elecktrons. *Zeitschrift für Physik*, 48:11–14, 1928.

[4] A. Einstein, B. Podolsky, and N. Rosen. Can quantum-mechanical description of physical reality be considered complete. *Phys. Rev.*, 47:777, 1935.

[5] P. A. M. Dirac. *The Principles of Quantum Mechanics*. Oxford Unv. Press, London, 4nd edition, 1958.

[6] D. N. Schwartz. *The Last Man Who Knew Evergything*. Basic Books, NYC, 2017.

[7] D. P. Monslesan and et.al. On the archetypal "flavors", indices and teleconnections of enso revealed by global sea surface temperature, arxiv:2406.08694v1 [physics,ao-ph], 2024.

[8] K. Ferguson. *Lost Science*. Union Square & Co., NYC, 2017.

[9] H. Rowland. Magnetic permeability and the maximum of magnetism of iron, steel and nickel. *Phil. Mag.*, 46:140–150, 1873.

[10] J.P. Wallace and M.J. Wallace. *"yes Virginia, Quantum Mechanics can be Understood" 3rd. ed.* Casting Analysis Corp., Weyers Cave, VA, 2025.

[11] J. Barzun. *The House of Intellect*. Harper & Brothers, NYC, 1959.

[12] D. Higinbotham. private comm. the threshold for generating the nuclear short range correlation in deuterium is with an electon having an energy between 4-5 gev that can tranfer ~ 1 gev to merge the nucelons into a high energy state that promptly decays into a proton and neturon moving appart at high energy in opposite directions., 2014.

[13] J. P. Wallace and M. J. Wallace. The bound state, http://vixra.org/pdf/2103.0026v2.pdf, 2024.

[14] A. Pickering. *Constructing Quarks*. Uni. Chicago Press, Chicago, 1984.

[15] S. Raimes. *The Wave Mechanics of Electrons in Metals*. North-Holland Publishing Company, Amsterdam, 1961.

[16] J.P. Wallace and M.J. Wallace. *The Principles of Matter amending quantum mechanics*. Casting Analysis Corp., Weyers Cave, VA, 2014.

[17] J.P. Wallace and M.J. Wallace. *"yes Virginia, Quantum Mechanics can be Understood"*. Casting Analysis Corp., Weyers Cave, VA, 2017.

[18] P. Ehernfest. Bemerkung über die angenäherte gültigkeit der klassischen mechanik innerhalb der quantenmechanik. *Zeits. f. Physik*, 45:455–457, 1927.

[19] S. Chandrasekhar. *Newton's Principia for the Common Reader*. Clarendon Press, Oxford, 1995.

[20] M.C. Gutzwiller. *Chaos in Classical and Quantum Mechanics*. Springer-Verlag, NYC, 1990.

[21] P.A.M. Dirac. Forms of relativistic dynamics. *Rev. Mod. Phys.*, 21(3):392–399, 1949.

[22] J.P. Wallace and M.J. Wallace. *"yes Virginia, Quantum Mechanics can be Understood"* 2nd ed. Casting Analysis Corp., Weyers Cave, VA, 2020.

[23] J.P. Wallace. Electrodynamics in iron and steel, arxiv:0901.1631v2 [physics.gen-ph], 2009.

[24] J.P. Wallace. Spintronics enter the iron age. *JOM*, 61(6):67–71, June 2009.

[25] F. Brailsford. *Physical Principles of Magnetism*. van Nostrand, London, 1966.

[26] J.P. Wallace and M.J. Wallace. Relativistic longitudinal spin wave. *http://vixra.org/pdf/1405.0015v1.pdf*, 2014.

[27] J.P. Wallace. Proton in srf niobium. In G.R. Myneni and et. al., editors, *SSTIN10 AIP Conference Proceedings 1352*, pages 205–312, Melville, NY, 2011. AIP.

[28] P. A. M. Dirac. Relativistic quantum mechanics. *Proc.Roy. Soc. A*, 136:453–464, 1932.

[29] James C. Maxwell. *A Treatise on Electricity and Magnetism Vol II.* Dover Press, NYC, 3rd 1891 edition, 1866. reproduction.

[30] J. D. Jackson. *Classical Electrodynamics.* John Wiley & Sons, NYC, 2nd edition, 1975.

[31] M. Kac. Can one hear the shape of a drum. *American Mathematical Monthly*, 73(4 part 2):73, 1966.

[32] J. P. Wallace and M. J. Wallace. 21cm quantum amplifier, http://vixra.org/pdf/2204.0024v3.pdf, 2022.

[33] J. v. Neumann. *Mathematical Foundations of Quantum Mechanics.* Princeton Univ. Press, org 1932 trans 1953. trans. Beyer, R.T.

[34] D. Ferry. *The Copenhagen Conspiracy.* Pan Stanford Pub., Singapore, 2019.

[35] J.P. Wallace and M.J. Wallace. Electrostatics. In G.R. Myneni, editor, *Science and Technology of Ingot Niobium for Superconducting Radio Frequency Applications*, volume 1687, pages 040004–1–14, Melville, NY, 2015. AIP. *Electrostatics.*

[36] E. Fermi. Quantum theory of radiation. *Rev. Mod. Phys.*, 4:87–132, 1932.

[37] P.M.A. Dirac. Quantized singularities in the electromagnetic field. *Proc. Roy. Soc. A*, 133(821):60–72, 1931.

[38] M. Gell-Mann and F.E. Low. Quantum electrodynamics at small distances. *Phys. Rev.*, 95(5):1300–1312, 1954.

[39] R. Penrsoe. *The Road to Reality.* Vintage Press, NYC, 2004.

[40] W.C. Lamb and R.C. Retherford. Fine structure of the hydrogen atom by a microwave method. *Phys. Rev.*, 72(3):241–243, 1947.

[41] H.M. Foley and P. Kusch. On the intrinsic moment of the electron. *Phys. Rev.*, 73:412, 1948.

[42] W. Heisenberg. *The Physical Principles of the Quantum Theory.* Unvi. Chicago, Chicago, 1930.

[43] R.P. Feynman, R.B. Morinigo, and W.G. Wagner. *Feynman Lectures on Gravitation.* Westview Press, Bolder Co., 2003.

[44] S. Weinberg. *The Quantum Theory of Fields Vol I and II.* Cambridge Unvi. Press, Cambridge, 1995.

[45] J.P. Wallace and M.J. Wallace. Refraction, http://vixra.org/pdf/1809.0582v2.pdf, 2018.

[46] J.P. Wallace and M.J. Wallace. *Dark Matter from Light extending quantum mechanics to Newton's First Law.* Casting Analysis Corp., Weyers Cave, VA, 2011.

[47] A. Pais. *Subtle is the Lord.* Oxford Univ. Press, NYC, 1982.

[48] P. Kusch. Summary remarks on research institutes within universities. Personal Papers Press Release from Columbia Univ., March 1962. Spring Meeting in Wash. DC of APS.

[49] J. W. Dauben. *George Cantor His Mathematics and Philosophy of the Infinite.* Harvard Univ. Press, Cambridge, MA, 1979.

[50] J. Schwinger. *Einstein's Legacy.* Scientific American Books, NYC, 1986.

[51] A. Pais. *Inward Bound.* Oxford Univ. Press, NYC, 1986.

[52] M. Kac. *Statistical Independence in Probability, Analysis and Number Theory.* #12 The Carus Mathematical Monographs. The Math. Assoc. of America, Rahway, NJ, 1959.

[53] C. W. Misner, K. S. Thorne, and J. A. Wheeler. *Gravitation.* W. H. Freeman and Co., San Francisco, 1972.

[54] S. Weinberg. *Gravitation and Cosmology: Principles and Applications of the General Theory of Relativity.* John Wiley & Sons, NYC, 1972.

[55] B. F. Schutz. *A First Course in General Relativity*. Cambridge Univ. Press, Cambridge, 1985.

[56] J Bell. On the einstein, podolsky, rosen paradox. *Physics*, 1(3):195–200, 1964.

[57] A. D. Aczel. *Entanglement*. Plume Penguin, NYC, 2003.

[58] J.P. Wallace and M.J. Wallace. Statistical independence in quantum mechanics. *J. Phys. Astro.*, 11(4):337, 2023. https://vixra.org/pdf/2303.0047v3.pdf.

[59] H. Everett. *Theory of the Universal Wavefunction*. PhD thesis, Princeton Univ., 1956.

[60] J. P. Wallace and M. J. Wallace. The big bang's quantum problem, http://vixra.org/pdf/2107.0123v3.pdf, 2021.

[61] J. A. Wheeler. *Geons Black Holes Quantum Foam*. Norton Press, NYC, 1998.

[62] P.C.W. Davies and J.R. Brown, editors. *The Ghost in the Atom*. Cambridge Univ. Press, Cambridge, 1986.

[63] R Aris. Catastrophe theory. pre-print, 1978. remarks at the end of the lecture.

[64] H.A. Bethe and E.S. Salpeter. *Quantum Mechanics of One- and Two-Electron Atoms*. Springer, Berlin, 1957.

[65] C. Moore. *Atomic Energy Level*, volume 1. NBS, Washington, DC, 1971. NBS-PUB-C 197.

[66] J. P. Wallace and M. J. Wallace. Nucleons, https://vixra.org/pdf/2506.0155v1.pdf, 2025.

[67] J. P. Wallace and M. J. Wallace. The origin of the meissner effect and superconductivity, http://vixra.org/pdf/2204.0024v2.pdf, 2024.

[68] S. Chandrasekhar and E. Fermi. Problems of gravitational stability in the presence of a magnetic field. *Astrophys. J.*, 118:116–141, 1953.

[69] E. Fermi. *Nuclear Physics*. University of Chicago Press, Chicago, Ill., 1949. edt. Orear, J. and Rosenfeld, A. H. and Schluter, R.A.

[70] G. Gamow. Gravity. *Scientific American*, (3), 1961.

[71] P. Kusch. Summary remarks on teaching. Personal Papers Press Release from Columbia Univ., Feb. 1960. Meeting in Detroit MI.

[72] S. S. Schweber. *QED and the Men Who Made It: Dyson, Feynman, Schwinger, and Tomonaga*. Princeton Unvi. Press, Princeton, N.J., 1994.

[73] L. J. Slater. Confluent hypergeometric functions. In M. Abramowitz and I.A. Stegun, editors, *Handbook of Mathematical Functions with Formulas, Graphs, and Mathematical Tables*, ASM 55, pages 503–536. Dept. of Commerce, Washington DC, 1968.

[74] W. Pauli and V. Weisskopf. The quantization of the scaler relativistic wave equation. *Helv. Phys. Acta*, 7:709, 1934.

[75] G. Feinberg. Possibility of faster-than-light particles. *Phys. Rev.*, 159:1089, 1967.

[76] J.A. Formaggio and G.P. Zeller. From ev to gev: Neutrino cross-sections across energy scales. *Rev. Mod. Phys.*, 84:1307–1354, 2012. arXiv: 1305.7513v1.

www.ingramcontent.com/pod-product-compliance
Lightning Source LLC
Chambersburg PA
CBHW021934190326
41519CB00009B/1022